ECOHYDROLOGY OF THE ANDES PARAMO REGION

Cover page: Fully attributed to Guido Chavez.

ECOHYDROLOGY OF THE ANDES PARAMO REGION

DISSERTATION

Submitted in fulfillment of the requirements of
the Board for Doctorates of Delft University of Technology
and
of the Academic Board of the UNESCO-IHE
Institute for Water Education
for
the Degree of DOCTOR
to be defended in public on
20, December 2016, 12:30 PM
In Delft, the Netherlands

by

Veronica Graciela MINAYA MALDONADO

Master in Hydric Resources and Water Science, EPN, Quito - Ecuador
Master of Science in Environmental Science, UNESCO-IHE, Delft - The Netherlands

born in Quito, Ecuador

This dissertation has been approved by the
promotor: Prof.dr.ir. A. E. Mynett

Composition of the Doctoral Committee:

Chairman	Rector Magnificus, Delft University of Technology
Vice-Chairman	Rector UNESCO-IHE
Prof.dr.ir. A. E. Mynett	UNESCO-IHE / Delft University of Technology, promotor
Dr. J. van der Kwast	UNESCO-IHE

Independent members:

Prof.dr.ir. H. H. G. Savenije	Delft University of Technology
Prof.dr. M. E. McClain	UNESCO-IHE / Delft University of Technology
Prof.dr. W. Buytaert	Imperial College London
Prof.dr.ir. R. Galárraga	Escuela Politécnica Nacional del Ecuador
Prof.dr.ir. C. Zevenbergen	Delft University of Technology / UNESCO-IHE, reserve member

This research was conducted under the auspices of the Graduate School for Socio-Economic and Natural Sciences of the Environment (SENSE)

CRC Press/Balkema is an imprint of the Taylor & Francis Group, an informa business

Published by:
CRC Press/Balkema
PO Box 11320, 2301 EH Leiden, The Netherlands
e-mail: Pub.NL@taylorandfrancis.com
www.crcpress.com – www.taylorandfrancis.com

ISBN 978-1-138-63312-4. (Taylor & Francis Group)

Acknowledgments

I would like to express my sincere gratitude to Prof. Arthur Mynett for his understanding spirit and endless support either morally, financially or academically throughout my PhD. He gave me trust and built up my confidence to face challenges. I will ever be in debt with him.

This research could not have been completed without the effort and cooperation from my mentor Dr. Hans van der Kwast. I want to thank to Dr. Gerald Corzo for his guidance, support, interesting discussions and encouragement in finishing this research. Thanks to the external committee for their positive feedback; their contributions are sincerely appreciated and gratefully acknowledged.

Thanks to an important part of my life here in Delft, my IHE family and friends I met along the way and whom I will always be grateful for. Pato!, you are one of a kind, thanks for all these years of friendship; you are in many ways my example to follow. Many thanks to Ana, Adey, Adris, Aky, Alida, Aline, Andres, Angy, Arlex, Arvind, Benno, Carlitos, Cesar, Fer, Gaby, Jessy, July, Marianne, Maribel, Mauri, Miguel, Mohan, Mohaned, Nata, Neiler, Pin, Sayra, Vivi, Yared, and Zahra. I always suspect that Juanca was one of those super smart aliens that came to earth to help humans like me; thanks for those fruitful conversations about hydrology and life.

How can I forget my "Water Youth Network", an amazing team of truly young water leaders. Your enthusiasm, helpfulness and dedication to empower youth is very contagious!. I have learnt so much from all of you; special thanks to Alix, Ceci, Dona, Janet, Laura, Maelis, Robert, Roos, Shabana, Vero D, and many other energetic souls within the network.

My friends from Ecuador, whom I have a very strong connection with; Alexandra, Aly, Andyman, Andre G, Cyntia, Flaka, Luis, Mario, Mariela, Monica, Omar, Rafael and Tere. A special thanks to Erika, more than my friend you are a sister to me. Every time we meet feels like we have never been apart; we enjoy and laugh like the old times.

Thousand thanks to my dearest family, my father Manuel, my siblings Alejo, Alex, Cyntia, Francis, Roberto, Sammy and my aunt Vicky. Words cannot express the gratitude and love I have for all of you; from far I felt you very close to me. Thanks to all the "Van der Steen" family in special to Harry, Linda, Ernst and Vivian for making me feel part of you and provide me a home away from home. Last but not least to my liefje Mark for his patience, support and love without conditions.

Para mi familia, este libro y toda mi vida.

Summary

Tropical grasslands are one of the most abundant but probably least understood ecosystems in terms of their biological and physical processes. In the Andean region such grasslands are known as *páramos,* and although widely recognized for sustaining biodiversity, carbon sequestration and water storage, the *páramos* have become vulnerable not only to climate change but also to land use change due to e.g. agriculture, grazing and burning. Usually these changes are associated with socio-economic factors driving communities to aspire higher income generation. Currently there are some initiatives aiming to protect and conserve the *páramos* ecosystems by improving living conditions for nearby communities and compensating for ecosystem services. These initiatives even include a payment strategy to slow down and potentially stop land conversion. However, the current procedure does not take into account any quantified methodology to assess the difference in providing ecosystem services.

In the Ecuadorian *páramos,* recent studies focused on obtaining a better understanding in specific areas, e.g. plant taxonomy, biodiversity, hydrology, among others. However, most of them are limited to individual concepts of specific research areas without including a comprehensive ecosystem analysis that includes adequate assessment of ecosystem services that the *páramos* provide. The use of an integrated approach including field experiments and numerical modelling to explore the behavioural components of the *páramos* ecosystems is not common, mainly because of the complex processes interactions that unfortunately are still not well understood in these types of regions. Current modelling approaches often do not contemplate the altitudinal variation of ecological processes at different elevations and for different vegetation types in regions like the *páramos,* mainly due to data unavailability. Likewise, relevant runoff processes are often not well understood in the Andean Region due to the high spatial variability of precipitation, the properties of young volcanic ash soils, the soil moisture dynamics and other local factors such as vegetation interception and high radiation that might influence the hydrological behaviour. Thus, any information available is often extrapolated to unsampled areas with a high degree of uncertainty and without considering the particular carbon and nitrogen composition of vegetation, soil characteristics and quantification of runoff components along altitudinal ranges.

The study area considered in this thesis was a pilot region located within the Antisana Ecological Reserve (628.1 km^2) in the Andean Region of Ecuador. The area is highly important since it is one of the main water sources for La Mica Reservoir, which supplies water and generates electricity for more than half million inhabitants in southern Quito, Ecuador. This *páramo* ecosystem supplies important environmental services. The aim of this research is two-fold,: (1) To contribute to the understanding of the interactions between processes of different nature through the integration of field experiments and modelling techniques that represent the functioning of the Andean *páramos*; and (2) To propose a series of environmental services indicators to quantify the regulation and maintenance services provided by the *páramo* ecosystem. The intention is to explore the processes and interactions within the *páramo* ecosystem and to contribute to a better quantification of ecosystem services, strengthening the sustainability of integrated management strategies of these high–altitudinal regions.

To achieve this, a number of studies were carried out: (i) extensive fieldwork and statistical analysis to identify the differentiation of vegetation physiology and catchment characterization along an altitudinal gradient; (ii) selection and testing of a biogeochemical model (BIOME-BGC) for analysing gross primary production and hydrological processes; (iii) analysis of relationships between climatic variables and gross primary production; (iv) hydrochemical catchment characterization and quantification of runoff generation; (v) testing of the selected process-oriented hydrological model; (vi) identification and quantification of the ecosystem services provided by the *páramos*.

This integrated research started with a comprehensive fieldwork that assessed the main ecophysiological parameters that were not readily available from literature, identified the main growth forms of vegetation and quantified the carbon stocks currently available in the area. A parameterization of the main variables was used to validate and test the biogeochemical and ecophysiological model BIOME-BGC that was chosen among other similar models since its focus was on the gross primary production and the hydrological processes. The gross primary production and the hydrological budgets were estimated by taking into account the main properties of the *páramos ecosystems* such as plant functional types, site/soil parameters and daily meteorology. Also, key sensitivities in the soil-vegetation interaction components were identified.

A number of statistical data analysis and data driven models were used to evaluate promptly the complex relations between gross primary production and climatic variables that are relatively easy to measure. Here, short wave radiation, vapour pressure deficit, and temperature were the main drivers for gross primary production variation on a monthly basis. Surprisingly, the analysis showed that precipitation was not a variable that directly seemed to influence the variation of gross primary production; however, it is well known that precipitation is the major driving force for plant growth and therefore carbon uptake by plants. To better understand the overland flow contribution and how precipitation is interacting in the hydrological system, the runoff components of the catchment were investigated further. This was done using a spatial hydrochemical characterization of flow pathways and routing analysis. The findings were used in a process-realistic description of the runoff generation mechanisms described in several hydrological units. These are key elements of the process-oriented hydrological Tracer Aided Catchment model, the distributed (TAC^D) that was successfully applied in our case study.

The ecosystem services were assessed by using all information from previous studies on carbon stocks and water resources availability in the region. These were used to quantify the ecosystem services and build indicators for water regulation and carbon sequestration in the *páramo* ecosystem. The outcome of this study contributes to develop strategies and good management practices in the *páramo* ecosystem.

Despite all limitations in input data, hydrological process understanding, vegetation interaction, among others, the intention of this PhD research is that the present study provides a comprehensive framework that can be applied to understand the vegetation-soil-water-climate interactions in these combined glacier-*páramo* catchments in the Andes Region. The thesis also aims at providing adequate tools as a step towards a fair *páramo* ecosystem services assessment. Additionally, the approach developed in this thesis could be used to examine the response of the *páramos* to different scenarios by adding climate variability (e.g. el Niño phenomena), change of land cover, grazing, burning and by using the tools developed here to analyze the resilience of the *páramo* ecosystem and how this will affect the benefits from its ecosystem services.

Samenvatting

Tropische grasgebieden komen veel voor, maar hun ecosystemen worden nog nauwelijks begrepen waar het de onderliggende biologische en fysische processen betreft. In het Andes gebergte in Zuid Amerika staan deze grasgebieden bekend als páramos en hoewel hun belang voor het behoud van biodiversiteit, het opnemen van CO2, en het opslaan van water breed wordt onderkend, zijn deze páramos gebieden kwetsbaar voor bijv. klimaatverandering en veranderingen in landgebruik door intensivering van landbouw en veeteelt. Deze veranderingen hebben vaak te maken met socio-economische factoren die de lokale bevolking ertoe zetten om hun inkomens-positie te verbeteren. Er zijn momenteel zelfs initiatieven gaande om de páramos ecosystemen te beschermen en te behouden door de leefomstandigheden van lokale gemeenschappen te verbeteren en het eventuele verlies van ecosysteem functies te compenseren. Deze initiatieven gaan zelfs gepaard met financiële steun aan de plaatselijke bevolking om het verlies van páramos gebieden te vertragen en zo mogelijk te stoppen. De huidige aanpak is echter niet gestoeld op enige wetenschappelijke onderbouwing.

In de páramos gebieden van Ecuador is onlangs begonnen met onderzoek naar ondermeer de taxonomie van plantensoorten, biodiversiteit, hydrologische processen, etc. Echter, de meeste onderzoeken zijn beperkt tot detailaspecten van specifieke onderzoeksgebieden zonder dat aandacht wordt besteed aan een uitgebreide analyse van alle ecosysteemfuncties die de páramos gebieden leveren. Een integrale benadering gebaseerd op veldwaarnemingen en (numerieke) modelvorming is niet gebruikelijk, met name vanwege de complexe interacties tussen verschillende processen die nog steeds niet goed begrepen worden. Zo houden de huidige modelaanpakken veelal geen rekening met het effect van hoogteverschillen op vegetatiesoorten en ecologische processen, voornamelijk omdat er geen gegevens hierover beschikbaar zijn. Om dezelfde reden worden ook de hydrologische processen in het Andes gebergte niet goed begrepen vanwege de sterke ruimtelijke variatie in neerslag intensiteit, de specifieke eigenschappen van de vulkanische ondergrond, de dynamica van bodemvocht, en andere lokale factoren zoals de invloed van vegetatie op hydrologische processen.

Vandaar dat de beperkt beschikbare informatie vaak wordt toegepast in gebieden waar geen gegevens beschikbaar zijn, wat de nodige onzekerheid met zich meebrengt omtrent de specifieke koolstof en stikstof samenstelling van de aanwezige vegetatie, de lokale grondeigenschappen, en de hydrologische processen op verschillende hoogten. et studiegebied voor dit proefschrift was een testgebied gelegen in het Antisana Ecologisch Reservaat (628.1 km²) in het Andes gebergte van Ecuador. Dit gebied is een belangrijke toevoerbron van water naar het La Mica reservoir, wat water en elektriciteit levert voor meer dan een half miljoen inwoners van het zuidelijke deel van Quito. Dit *páramo* ecosysteem is dan ook van groot belang voor het gebied.

Het doel van dit proefschrift is tweeledig: (1) bijdragen aan het doorgronden van de interacties tussen de verschillende processen door gebruik te maken van veldonderzoek en (numerieke) modellering van het gedrag van de Andes *páramo*; en (2) om een aantal indicatoren te ontwikkelen die kunnen worden gebruikt om de toestand van het *páramo* ecosysteem goed weer te geven. Het streven is er op gericht om de relevante processen en hun interacties te onderzoeken en te begrijpen om zodoende een duurzame ontwikkeling en beheer van deze gebieden te kunnen aanbevelen.

Hiertoe is een aantal studies uitgevoerd: (i) uitgebreid veldonderzoek en statistische analyses van de meetgegevens teneinde het verschil in vegetatie op verschillende hoogten te kunnen vaststellen; (ii) het selecteren en testen van een biogeochemisch model (BIOME-BGC) voor het analyseren van de primaire productie en hydrologische processen; (iii) onderzoek naar de relatie tussen klimaat en primaire productie; (iv) vaststellen en kwantificeren van hydrochemische grootheden en de hydrologische afvoer van het onderzochte stroomgebied; (v) het testen van het geselecteerde proces-gerichte hydrologische model; (vi) het identificeren en kwantificeren van het belang van het *páramo* ecosysteem.

De integrale benadering die hier is gevolgd begon met uitgebreid veldonderzoek naar de belangrijkste ecofysiologische parameters van het gebied voor zover die niet op basis van literatuur onderzoek kon worden verkregen. Daarbij zijn groeivormen van de vegetatie vastgesteld alsmede de hoeveelheid gebonden koolstof in het gebied. Op basis hiervan zijn de belangrijkste parameters gebruikt om het biochemische en ecofysiologisch model BIOME-BGC te valideren en te testen. Dit model is gekozen omdat het specifiek gericht is op primaire productie en hydrologische processen. Door gebruik te maken van de specifieke plant/bodem eigenschappen van de (deel)stroomgebieden in combinatie met dagelijkse meteorologische omstandigheden, konden met dit model schattingen van de primaire productie en de waterbalans worden gemaakt. Bovendien kon de gevoeligheid van parameters op de interactieprocessen worden vastgesteld.

Gebruik makend van verschillende statistische en data-gedreven modellen konden de complexe interacties tussen primaire productie en klimaatgrootheden worden afgeschat. Op basis van parameters als uitstraling, vochtgehalte en temperatuur werd de maandelijkse variatie in primaire productie ingeschat. Daarbij bleek, enigszins onverwacht, dat niet zozeer de neerslag de belangrijkste factor is voor primaire productie, hoewel bekend is dat dit wel de belangrijkste factor is voor plantgroei en dus voor opname van koolstof. Om beter te begrijpen wat het belang is van afstroming van regenval en hoe dit het hydrologisch systeem beïnvloedt, is nader onderzoek uitgevoerd. Daartoe zijn isotopen uitgezet en stroombanen bepaald. De resultaten hiervan zijn gebruikt om een realistische beschrijving op te stellen van de processen die in de verschillende (deel)stroomgebieden spelen. Deze werden vervolgens gebruikt in het hydrologisch proces-model TAC^D (Tracer Aided Catchment model).

Op basis van alle deelstudies met betrekking tot CO_2 opslag en (drink)watervoorziening voor de regio is uiteindelijk een model opgesteld waarmee het belang van het *páramo* ecosysteem kon worden nagegaan. Op basis hiervan werden indicatoren ontwikkeld die vervolgens kunnen worden gebruikt om het effect van maatregelen en de duurzaamheid van het beheer van het *páramo* ecosysteem vast te stellen.

Ondanks alle beperkingen van beschikbaarheid van gegevens, begrip van de processen en hun interacties, beperkingen in modelvorming, etc. bestaat de overtuiging dat dit proefschrift heeft bijgedragen aan de ontwikkeling van een breed raamwerk waarmee de interacties tussen vegetatie-bodem-water-klimaat kunnen worden ingeschat voor hoger gelegen *gletsjer–páramo* stroomgebieden in het Andes gebergte. Het doel van dit proefschrift is ook om instrumenten te bieden waarmee de gevolgen van veranderingen in het *páramo* ecosysteem kunnen worden ingeschat. Met behulp hiervan kunnen verschillende scenario's worden onderzocht naar effecten van klimaatverandering (zoals el Niño), veranderingen in landgebruik, intensivering van landbouw en veeteelt, etc. Op die manier kan de veerkracht van de Andes *páramo* worden bepaald en het effect van maatregelen worden nagaan om de belangrijke functies als CO_2 opslag en watervoorziening in deze waardevolle gebieden te behouden.

Contents

If I had one hour to save the world I would spent 55 minutes
defining the problem and only 5 minutes finding the solution
(A. Einstein)

GENERAL INTRODUCTION

1.1 Background

High altitudinal ecosystems are a complex interaction of physical, biotic, abiotic and anthropogenic factors (Azocar and Rada, 2006; Cuatrecasas, 1979; Ricardi et al., 1997; Sklenar and Jørgensen, 1999; Vargas et al., 2002). The complex spatial, biological and physical patterns of these ecosystems are related to key insight processes and many interrelated factors such as: altitude, disturbance and the availability of safe sites at higher altitudes (Cavalier, 1996; Hilt and Fiedler, 2005; Küper et al., 2004; Sklenar and Ramsay, 2001). The tropical region of northern South America holds different types of ecosystems that are characterized by a specific type of vegetation also known as *páramos*, which are located at elevations between 3000 and 4700 m a.s.l. The *páramos* are mainly high tropical montaine vegetation (Lauer, 1981; Monasterio and Sarmiento, 1991; Walter, 1973) that are discontinuously distributed between 11°N and 8°S latitudes (Luteyn, 1999). However, many other studies (Brack Egg, 1986; Cleef, 1978; Jørgensen and Ulloa, 1994; Monasterio, 1980; Ramsay, 1992; Vuilleumier and Monasterio, 1986) have different geographical locations for *páramos*, especially for neotropical areas that have *páramo*-like vegetation (Luteyn, 1999). The *páramos* in Ecuador are influenced by intertropical converge air masses and throughout the year are very humid, they receive more than 2000 mm of rain per year (Hofstede et al., 2002; Luteyn, 1999).

The *páramos* provide important environmental services to both local and global scales, for instance, the tropical andean ecosystems in South America are well known by providing extremely important services such as biodiversity conservation, carbon storage, water supply and regulation (Buytaert et al., 2011; Myers et al., 2000). If we look at the Ecuadorian *páramos* more closely, the biodiversity holds unique fauna and flora that have adapted to the particular climatic conditions of the region, which has been found to have up to 60% of endemic species (not found anywhere in the world). One of the most important features of the soils of *páramo* regions is that these act as carbon sinks, where the organic carbon basically is stored and accumulated due to the formation of resistant organometallic complexes typical of volcanic material (Shoji et al., 1993). The particularly striking aspect of the *páramos* is its orographic properties; these high altitudinal ecosystems receive higher amounts of precipitation, thus having a good capacity for water regulation and storage. The *páramos* hold and release the water gradually during summer (Greiber and Schiele, 2011) and trap surplus water during rainy seasons, thus controlling the level of water in the Andean Rivers. These ecosystems are considered crucial for local water supply for cities, agriculture and hydro-power and they constitute the only source of water for the population in the upper and lower part of the Andes (Greiber and Schiele, 2011). Thoumi and Hofstede (2012) pointed out the great ecological significance that the *páramo* represents to the Ecuadorian population, rating it up to 90% of importance. The *páramo* landscape has been influenced by glaciations and therefore it is difficult to give a single definition due to its diverse geographic, geologic, climatic and floristic features (Luteyn, 1999). It is mainly composed by grasslands (scrublands, pasture lands and meadows) with a large variety of endemic floral species (Bosman et al., 1993). The most known and main species in the *páramo* are postrate schrubs (*Loricaria*), cushions (*Plantago rigida*), aculescent rosettes (*Hypochaeris*) and tussock grasses (*Festuca*) (Cuesta and De Bievre, 2008). The *páramo* ecosystems exhibit specific governing features such as low temperatures, high humidity, soils with a high content of organic matter, low phosphorus availability and acid pH (Hofstede et al., 2002; Tonneijck et al., 2010).

At higher elevations, mostly above 4500 m a.sl., the *páramo* vegetation become more sparse surrounded by moraines and glaciers. The Andean glaciers have demonstrated to be correlated to temperature fluctuations, these glaciers have changed radically in the last 50 years (Francou et al., 2000) losing more than 40% of the area in a short period of time (Cadier et al., 2007). The glaciers in the highlands are the most sensitive indicators to the climate change trends, several studies point that the total melt of the glaciers are expected between 2020-2030 (Cadier et al., 2007; Francou, 2007; Francou et al., 2007; Marengo et al., 2010). In this context the study of the *páramo* vegetation, which at certain extent depends on the glacier, is a priority. A comprehensive study of the *páramo*, its structure and functionality will help to explore and understand the impacts that climate change might have on these ecosystems.

The moraine is located from about 4600 to 5000 m a.s.l. (Luteyn, 1999) and it is characterized by the lowest air temperature, precipitation, water retention in soil and nutrient content (Baruch, 1984). This type of ecosystem is the less disturbed in terms of human intervention and it holds a variety of some scattered growing vegetation on rocks and sand soil typical from that specific vegetation zone. The vegetation here depends not only on site-specific water availability, which is highly dependent on the precipitation pattern (Bosman et al., 1993), but also on the soil texture and nutrient availability. Basically this ecosystem is a product from the retreatment of the glacier so it can be considered as sediments placed from the glacier movement. The moraine ecosystem is subjected to cyclic processes due to the ice melting and the prevailing meteorological conditions at that specific altitude. Jorgensen & Ulloa

(1994) named this type of ecosystem as desert *páramo* where the vegetation is scarce due to the low temperatures and soil erosion.

1.1.1 Ecosystem and environmental interactions

Vegetation communities located in high-altitudinal ecosystems deal with many environmental factors (sunlight, temperature, carbon dioxide, altitude, soil and hydrology), which are essential components of the ecosystem-atmosphere interaction (Lange et al., 1998) (Figure 1-1).

In the *páramos*, the soil and vegetation history in some way is quite recently as a result of the last glacial period, when *páramos* were covered by glaciers (Hansen et al., 2003; Rodbell et al., 2002). The impacts of the climate change on the geographical distribution of the *páramo* vegetation will be strongly altered by changes in temperature and humidity (Buytaert et al., 2011). The spatio-temporal changes in the precipitation may have a great impact on the soil formation and ecosystem dynamics. In this regard, the displacement and expansion of sparsely vegetated areas is a potential indication of changes in the climate variability, which in turn is linked to the atmospheric interactions.

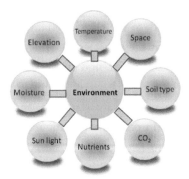

Figure 1-1 Soil-vegetation and environmental interaction.

Soil temperature has a deeper effect on nutrient and water availability, root growth and vegetation zonation (Diemer, 1996) and it is colder than the air temperature in a range between 0.4°C and 1°C (Cavalier, 1996). The reduction in temperature is reflected in the size of the vegetation (Cuatrecasas, 1958), change in the plant morphology (Odland, 2009), reduction of the capacity for biomass production (Roderstein et al., 2005). The frostbite of the water in the soil produces hydrological stress during the night and early in the morning. These low temperatures in the soil produce physiological drought and become a limiting factor not only for the movement of the water through the roots but also for the growth of the plants (Azocar and Rada, 2006). A seasonal hydrological stress is also observed during the dry season, the high diurnal radiation causes a higher evaporation demand and therefore low availability of water in the soil (Azocar and Rada, 2006). Other type of stresses are also considered; nutritional stress due to the restricted nutrient availability in the soils, mechanical stress owing to the winds and the frost and defrost processes in the soil and energetic stress for the high cloudiness during wet season that restricts the entry of radiation needed for the photosynthesis (Monasterio and Sarmiento, 1991). The high species endemism is due to their strong adaptation to the specific physic-chemical and climatic conditions at high altitudes where low atmospheric pressure, air density, CO_2 partial pressure, O_2, water vapour, jointly with strong wind and high ultraviolet radiation constitute ecological restrictions for plants and animals (Buytaert et al., 2011).

1.1.2 *Páramo* ecosystem services

The services and economic values the *páramos* provide are recognized for the management and conservation of the *páramo* grasslands in Ecuador. Management and conservation have to deal on one hand with the biodiversity and hydrological regulation and on the other hand with the value-based livestock production. It is important to highlight that more than 500.000 people, most of them from indigenous communities, live near by the *páramos* and use them for productive agriculture (Greiber and Schiele, 2011).

There are, however threats on these environmental services, for instance: agrochemicals, quarries, roads, water reservoirs, pine cultivation to protect the hillsides around reservoirs against erosion, cultivation, intensive livestock grazing and fire. Possibly the most threatening for the ecosystem is the human pressure that has increased the agricultural boundary (Dercon et al., 1998), which in turn changes the carbon storage in terms of the level of erosion or fertilization and manuring. Grazing and burning were found to cause a large impact to the vegetation and soil at the local scale, reducing interception and transpiration and increasing runoff (Ataroff and Rada, 2000; Hamza and Anderson, 2005; Harden, 2006; Molina et al., 2007; Pizarro et al., 2006).

The land use types currently found in the lower zones of the catchment are small-scale farming and cattle grazing, in consequence the above and below carbon storage is very low and the soil exhibits signs of erosion (Dercon et al., 2007). All these human activities lead to a loss of biodiversity, reduction of storage capacity and soil carbon, and hydrological regulation of the ecosystems in the Andean region (Buytaert et al., 2006a; Celleri and Feyen, 2009; Sarmiento and Bottner, 2002; Yimer et al., 2007). Many studies in the northern *páramos* in Ecuador have concluded that the removal of the original vegetation will lead to irreversible degradation of the soil structure, loss of organic carbon storage, reduction of water storage and regulation capacity (Podwojewski et al., 2002; Poulenard et al., 2001; Verweij, 1995); however is still very much an unknown quantity. The degradation in other midmountain areas can be partially restored with natural vegetation, while the *páramo* soil degradation is considered as irreversible (Celleri and Feyen, 2009). According to Farley (2007) the placing of pine plantation has changed the highland landscapes in the last four decades. This change in the land use, mainly composed by alien species to increase the forest cover, has been adapted as a mitigation measure to reduce agricultural land; however, it may not necessarily bring environmental benefits.

Likewise, the degradation of these ecosystems might be triggered by the changes in the climatic drivers, which threaten the ecosystemic functions and environmental services further downstream. Climate change seems to displace the ecosystem boundaries, by strongly shortening the glacier, the displacement of the ice cover will reduce the water availability in a long term and cause high sediment loads in the streams due to the erosion in the places where ice has been retreated. There is a large uncertainty in the different predictions based on global circulation models; however, there is an undeniable trend leading to the increase of temperature (Arnell, 1999; IPCC., 2007; Still et al., 1999). For instance, the projected temperature variation has a mean increase of about $3 \pm 1.5°C$ over the Andes (Urrutia and Vuille, 2009). Urrutia & Vuille (2009) assured that this prediction will be noticeable in a shorter period of time in higher altitudes. Particularly, the precipitation is more variable and it is expected to result in higher precipitation intensities and longer dry seasons (Buytaert and Beven, 2009). The increase of temperature will induce warmer soil conditions, and changes in the hydrological regime which will modify the hydrological regulation and organic carbon storage (Buytaert et al., 2011). In small scale catchments, temporal and spatial variability may still be controlled by natural fluctuations in the climate such as El Niño phenomena (Buytaert et al., 2011), which might affect the local system dynamics and processes of these tropical Andean ecosystems. However, the uncertainty for future predictions remain associated to the lack of data and limitations in the use of regional hydrological models that can capture the high spatial variability of meteorological variables and heterogeneity of the region (Buytaert et al., 2006a).

Several studies have predicted that biodiversity will stress or perish, especially species that are sensitive and not able to adapt to the new environmental conditions, high reduction in the soil carbon storage, reduction of water production, erosion, extinction of the genetic resources, among others (Buytaert et al., 2011). The dynamics of the terrestrial ecosystems are function of the level of disturbances that take place in a wide spatio-temporal range (White and Pickett, 1985), mainly the interaction of the climatic, topographic, geomorphologic patterns and the dynamic structure of the vegetation (Huston, 1994). In this context, the lack of protection to these types of ecosystems in the Andean highlands might worsen some of the degradation processes that are taking place due to deforestation, burning and land use – conflicts. Buytaert et al. (2011) stated that a net carbon might be released to the atmosphere if the below-ground organic carbon storage is reduced in this tropical Andean region bringing even more negative consequences. However, it is still challenging to quantify the impacts due to the high variability of climatic drivers and high heterogeneity of vegetation and soil properties (Buytaert et al., 2011).

1.2 Motivation of the study

The tropical Ecuadorian highlands are essential ecosystems that sustain biodiversity, biological processes, carbon sequestration, and water storage and provision. The *páramos* have been recognized essential as source of water that work as sponges absorbing and storing large amounts of freshwater, which are released later during dry periods. Moreover they have an enormous capacity to store carbon in the soil as well as in the plant material. However, previous studies in the *páramo* carried out at different temporal and spatial scales have failed to consider the spatio–altitudinal variation of the complex water–soil– vegetation interactions and the high heterogeneity of the ecosystem. On top of that, the lack of evidence-based policy-making has compromised the protection and conservation initiatives. There is a constant increase of activities such as agriculture, grazing and burning associated with socio-economic factors from communities that aspire higher income generation.

1.3 General objective

The main objective of this research is to contribute to the understanding of the interactions and functioning of the Andean *páramos* as a step towards an effective ecosystem services assessment of these high–altitudinal ecosystems. The assessment comprises a realistic quantification of carbon capture and storage and a comprehensive analysis of the water resources.

1.3.1 Specific objectives

- To assess carbon and nitrogen concentrations in soil and vegetation, aboveground carbon stocks distribution and soil organic carbon stocks along an altitudinal gradient;
- To implement the selected biogeochemical and ecophysiological model to simulate carbon and water fluxes in the *páramo* ecosystem, test the model performance, in particular the gross primary production and water budget in the system;
- To analyze the relationship between climatic variables and the gross primary production using data–driven model techniques;
- To determine the origin and quantify the contribution of the main runoff components using environmental tracers (isotopes and major ions);
- To apply a process–oriented hydrological model that represents different runoff generations processes within the catchment;
- To assess the ecosystem services of the *páramos* based on key indicators of regulation & maintenance.

1.3.2 Research questions

The research questions are done for a small catchment within the Ecuadorian *páramo* region, which is extensively described in Chapter 2.

- How much is the altitudinal variation of physical and biological processes in the Antisana *páramo* region? ;
- To what extent are the existing biogeochemical models capable to evaluate the carbon and water fluxes in the *páramo* ecosystem?;
- Which are the climatic drivers that have a strong influence on the temporal and spatial variability of GPP?;
- What is the origin and how much is the contribution of runoff components in the *páramo* ecosystem?;
- Is it possible to determine the runoff generation processes in the catchment by applying a distributed conceptual model?;
- What are the main ecosystem services provided by the *páramo* and how can we quantify them?

This research has scientific and societal significance. Scientifically, it contributes to a better understanding of the ecohydrological processes in the system affected by high spatio–altitudinal variation of climatic drivers and high heterogeneity of soil and vegetation. This supports a better quantification of the ecosystem services of the *páramos* that will lead to a comprehensive assessment of the impacts that degradation and climate change might have on the water resources availability and carbon sequestration in these high–altitudinal Andean ecosystems. From a societal point of view, it contributes to policy actions aiming for conservation, wise use and restoration of these *páramo* ecosystems, these will benefit directly or indirectly to the community's well-being that rely on these catchments in the tropical regions.

1.4 Outline of the thesis

This thesis is structured in nine chapters that include the introduction and conclusions & recommendations. The body chapters (Chapters 3 to 8) follow the specific objectives presented above.

Chapter 2 presents the main information of the study area, including the location, vegetation characteristics, climatic data, soil and geological features.

Chapter 3 establishes relationships among soil, vegetation and altitude by characterizing the main predominant vegetation, carbon and nitrogen composition, soil texture and above and belowground carbon stocks and biomass along an altitudinal gradient.

Chapter 4 presents the results of a selected ecosystem processes model when applied to the *páramo* catchment using the ecophysiological characteristics determined in Chapter 3. The results of carbon and water fluxes are evaluated and limitations of the model are discussed.

Chapter 5 evaluates the relationship between the climatic variables and the gross primary production (GPP) using data-driven model techniques.

Chapter 6 presents an experimental field analysis of water isotopes and natural tracers to determine the contribution of the different runoff components during dry and wet conditions. The results are a spatial representativeness of the main runoff components.

Chapter 7 presents in detail the set-up of the process-oriented hydrological model and the definition of the different runoff generation. The model is further calibrated using a genetic algorithm that examines the flow simulations at the outlet of the catchment.

Chapter 8 evaluates the ecosystem services based on the key indicators of provisioning and regulation & maintenance of the *páramo* ecosystem, focusing in the water regulation and carbon sequestration services.

Chapter 9 synthesizes the main findings and proposes recommendations for future research.

Some detailed information is provided in Annexes in each chapter. The list of abbreviations together with a short biography and list of publications are at the end of the book.

"No zone of alpine vegetation in the temperate or cold parts of the globe can well be compared with that of the Páramos in the tropical Andes." "Nowhere, perhaps, can be found collected together, in so small a space, productions so beautiful, and so remarkable in regard to the geography of plants." (Alexander von Humboldt)

2

DESCRIPTION OF THE STUDY AREA

This chapter starts with a brief history of Alexander van Humboldt, who was the first scientist to record biological, geographical and meteorological data of the ecuadorian *páramos*. It contains a complete description of the study area putting the reader into context about the main features of these *páramo* ecosystems in terms of location, climate and vegetation traits. The information provided here is a combination of data acquisition from national governmental and non-governmental institutions, extensive literature review and fieldwork visits and campaigns. During the field recognisance survey, we identified a clear differentiation of vegetation physiology along an altitudinal gradient. In addition, the climate data collected also support the strong dependency with the altitude. In this regard, our analysis is based on altitude gradients of environmental factors and ecophysiological processes present in the *páramos*. In addition, we also describe the approach used to simplify the high diversity in species vegetation evidenced in the site.

2.1 History

The first scientific investigations in the Andes were carried out by Alexander van Humboldt (1769-1859), a German geographer, naturalist and explorer with extensive field work in North and South America, and Eastern Europe (Wilson, 1995). Van Humboldt was considered as the father of the Universal Modern Geography due to his quantitative work in the field of physical geography, plant geography and meteorology. He travelled in South America between 1799 and 1804 exploring and describing his findings from a modern scientific point of view (Bohn, 1853). In 1807 van Humboldt published an article on the "Geography of Plants" in which he described the flora and fauna found at each elevation in the ecuadorian *páramos* nearby the Chimborazo icecap (Zimmerer, 2011). His descriptions included detailed drawings, maps and information of the climatic conditions of temperature, humidity and atmospheric pressure, which have been use up to now. Alexander van Humboldt has been extensively recognized for his scientific work and his results of his Latin America expeditions (Walls, 2009). Figure 2-1a shows his former property located in the same catchment of this present study, where he spent time doing research. Figure 2-1b shows the Humboldt meteorological station located 150m from his former property. There are several geographical features, animal species, places, foundations, academic schools and research institutes named after him as recognition of his valuable contribution to science.

Figure 2-1 a) 'Hacienda Humboldt' former property of Alexander van Humboldt, and b) Humboldt meteorological station.

2.2 Location

The study area, the Los Crespos – Humboldt basin (15.2 km²), is situated within the Antisana Ecological Reserve (628.1 km²) in the Andean Region of Ecuador (4000-5300 m a.s.l.). It consists of 15% glacier, 68% *páramo* grassland and 17% moraine. The latter one is an ecosystem in transition between the *páramo* and the glacier. The basin's ecosystem is the main water source for La Mica Reservoir, which supplies water to more than half million inhabitants in southern Quito, Ecuador (Figure 2-2).

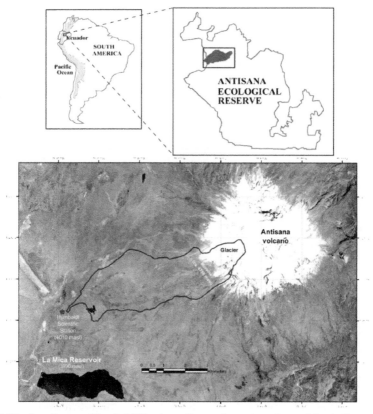

Figure 2-2 'The Los Crespos-Humboldt' basin on the south-western slope of the Antisana volcano in Ecuador. Source: ASTER Satellite Image, 15 m resolution.

2.3 Vegetation cover and soils

During a first recognisance survey in October 2012, a total of 9 growth forms were found at the site study, the only growth form absent according to the classification by Ramsay and Oxley (1997) was the Stem rosette, which can be only found in the *páramos* of northern Ecuador. Most of the plant species were identified at site to family level and in some cases to species level using identification field manuals (INEFAN., 1996; Jorgensen and Leon, 1999; Luteyn, 1999; Ronquillo., 2010) and the biologist expertise. In the lower and mid catchment there is a strong grass *páramo* dominance of tussock grasses (*Calamagrostis intermedia*), followed by acaulescent rosettes (*Werneria nubigena, Hypochaeris sessiliflora*), cushions (*Azorella Pedunculata*), postrate herbs (*Geranium multipartitum*) and postrate shrubs (*Bacchaeris caespitosa*)

(Figure 2-3). Bryophyta, fungi and lichens were recorded in most of the sampling sites with less than 2% coverage where other growth forms where abundant but at higher altitudes they could be seen in slightly higher numbers. The voluptuous tussock structure of the grass *páramo* below 4500 often exceeds 80% coverage, with some patch exceptions nearby flood zones and streams where other growth forms of vegetation were more dominant (cushions, erect herbs, acaulescent rosettes). Ramsay and Oxley (1997) highlighted the importance of tussocks, since they give shelter to other type of growth forms that under its covert could enlarge their size.

Figure 2-3 Growth forms of vegetation found in the study area, a) Tussocks (*Calamagrostis intermedia*), b) Acaulescent Rosettes (*Valeriana rigida*), c) Cushions (*Azorella pedunculata*), d) Postrate herbs (*Geranium multipartitum*), e) Postrate shrubs (*Baccharis caespitosa*), f) Upright shrubs (*Chuquiraga jussieui*)

During the field survey at 27 sampling sites (the detailed sampling design is described further in Chapter 3), we found that the highest number of taxa and individuals were located in the lower catchment (around 4000 to 4200 m a.s.l.). Conversely, the lowest numbers were found at higher altitudes as shown in Table 2-1.

Table 2-1 Summary table of the survey at the Los Crespos-Humboldt basin located in the south-western side of the Antisana volcano.

	Low (4000-4200 m a.s.l.)	Mid (4200-4400 m a.s.l.)	High (4400-4700 m a.s.l.)
# plots	10	10	7
# total individuals	61,651	50,859	22,216
individuals/m²	27.4	22.6	9.87
Shannon	3.077	3.219	3.084
# Families	27	25	20
# Genus	55	50	34
# Species	105	81	44
Top-3 growth-forms	TU,CU,AR	TU,AR,PS	AR,TU,PS

TU: tussock, CU: cushion, AR: acaulescent rosette, PS: prostrate shrub.

Our survey agreed with earlier studies (Ramsay and Oxley, 1997) to the fact that tussock, acaulescent rosettes, postrate herbs, and cushion are predominant growth forms in the area and in general in the Ecuadorian *páramos*, implying the strong relationship of the environmental settings. Altitudinal variation showed to be an important criteria as it is a proxy of the manifold drivers of the spatiotemporal dynamics of these ecosystems (Anthelme and Dangles, 2012; Körner, 2007; Nagy and Grabherr, 2009).

Soils are mainly andosols, based on the FAO classification (Gardi et al., 2014), derived from volcanic material characterized by their high soil moisture (Buytaert et al., 2005a) and water retention capacity (Janeau et al., 2015; Roa-García et al., 2011). In addition, studies in the area described an elevated amount of organic carbon and mineralogical composition in the soil of the *páramos*. The slopes are moderate (up to 15°) in the low and mid catchment and increases up to 30° close to the moraine at higher elevations. The glacier is an icecap that has retreated a couple of hundred meters in the last 20 years (Cáceres et al., 2005; Hall et al., 2012).

2.3.1 Vegetation traits

As stated earlier, the *páramo* vegetation is dominated by tussock grasses (TU), acaulescent rosettes (AR) and cushions (CU). The leaf longevity of the grasses is high but slightly declines with the altitude: TU longevity is 2.02 to 1.67 years; AR is 1.81 to 1.12 years; and CU is 1.30 to 0.98 years (Diemer, 1998). Due to low temperatures, growth rates are low and the leave's litter decomposition is slow (Spehn et al., 2006). The annual growth rate of young leaves is around 6.7 cm for TU (Scott, 1961), 40.2 cm for AR (Diemer, 1998), and 0.14 cm for CU. For CU, this translates to be 850 years (Ralp, 1978) in optimal conditions. The average height of leaves change significantly with elevation only for TU, while for AR and CU there is no significant difference

(Figure 2-4). The root length for TU is around 70 cm long in average for the low and mid altitudinal ranges, which helps to stabilize slopes and control erosion, whilst for AR the thick tap-roots vary around 15 cm long and for CU between 30 to 80 cm and in some cases can reach up to 100 cm long depending on the species and soil texture (Attenborough, 1995) (Figure 2-4).

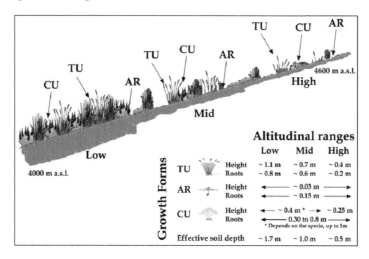

Figure 2-4 Height above ground and roots length for tussocks (TU), acaulescent rosettes (AR) and cushions (CU) at three altitudinal ranges (Low, Mid and High)

2.4 Climate

The area of the Antisana is representative for the eastern mountainous region and it is directly exposed to the humid wind of the Amazon River Basin (Manciati et al., 2011; Vuille et al., 2000). The north western slope of the glacier has a peculiar position, which makes it relatively protected from the wind and less cloudy. The zone exhibits local microclimates due to the strong fluctuation in temperature and precipitation due to the irregular topography and broadly differences in slope. The study area has a high relative humidity averaging 70-85% (Luteyn, 1999) and generally it has a cold and humid climate with some temperature fluctuations from below freezing to up to 30°C (Hedberg, 1964).

The Ecuadorian glaciers are essentially shaped by solid cones or geological structures along the Andean cordillera. Some of these snow and ice-capped mountains are active volcanoes. The glaciers show consecutive atmosphere states (precipitation, solar radiation, temperature, humidity, wind) by ice melting in their lower parts and they are key and very sensitive indicators of the climate variability (Francou and Vincent, 2007). The World Glacier Monitoring Service in Switzerland is the network in charge of the observation of the glaciers worldwide. The monitoring system organized by the program GreatIce involves 10 glaciers in the Andean tropics and provides information about the glaciers and hydro-meteorological data from the last 15 to 20 years.

2.5 Hydro - Meteorological data

From lower to higher altitudes the precipitation in the area varies from 900 to 1200 mm yr-1 and the average temperature from 7°C to 4.8 °C (for the period 2000 to 2010). Figure 2-5 shows the climate diagram of monthly averages values of precipitation and maximum and minimum temperatures at the two weather stations (Humboldt and Los Crespos Morrena) for the period 2000 to 2011. The wet period last typically from April to June. In the Ecuadorian Andes, the *páramos* above 3000 m a.s.l. receive 16% more precipitation compared to other *páramos* located in the inter-Andean valley (Buytaert and Beven, 2011). There are two sources of precipitation, one influenced from the air masses from the Amazon region and the second from the inter-Andean valley regime (Vuille et al., 2000).

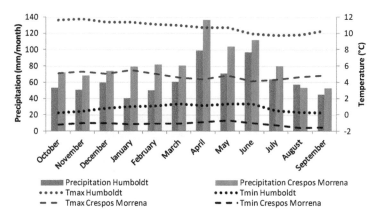

Figure 2-5 Average monthly precipitation, maximum and minimum temperatures at Humboldt and Los Crespos Morrena weather stations from 2000 to 2011

Precipitation has a large spatial variability (Buytaert et al., 2006a) with a presence of the so called "horizontal precipitation", which consist of fog and mist developed from the orographic uplift caused by the Andes (Buytaert et al., 2005b), which also limits transpiration (Bruijnzeel, 2004; Buytaert and Beven, 2011; Celleri and Feyen, 2009; Pizarro et al., 2006). Although, this additional source of water is minor and mostly intercepted by arbustive vegetation (Chuquiraga), other studies (Crockford and Richardson, 2000; Foot and Morgan, 2005) showed that the *páramo* ecosystem can catch low energy rain, drizzle and fog moisture on their leaves, which conduct over 50% of rainwater directly to the volcanic ash soils of Ecuadorian highlands (Janeau et al., 2015).

Figure 2-6 and Table 2-2 shows the hydro-meteorological stations located in the study area the type of information that is collected. These stations are maintained mainly by IRD (Institut de recherche pour le développement - Ecuador) and INAMHI (Instituto Nacional de Meteorología e Hidrología en Ecuador).

Table 2-2 Summary Hydro – meterological stations located in the Los Crespos - Humboldt basin

ID	Name	East	North	Elevation	Data
P6	Humboldt	810430	9943645	4059	Water Levels, precipitation, temperature, wind velocity and direction, conductivity (all data at 15 min frequency)
P7	Los Crespos	815067	9945705	4450	Water Levels
P8	Los Crespos Morrena	815834	9945610	4730	Precipitation, temperature, wind velocity and direction (all data at 15 min frequency)
P9	Páramo	812350	9946318	4269	Precipitation (monthly frequency)
P10	Camino de los Crespos	813175	9945200	4264	Precipitation (monthly frequency)

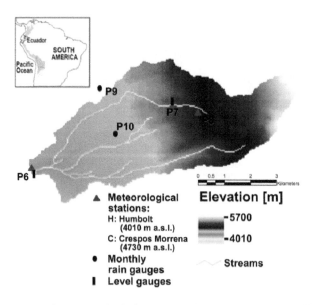

Figure 2-6 Hydro-meteorological stations in the Los Crespos-Humboldt basin

2.6 Geology

The geology of the catchment has a wide detric range that holds a variety of volcanic deposits from previous eruptions (Figure 2-7), the last significant eruption occurred nearly 1000 years ago based on stratigraphic studies (Hall et al., 2012). The peak is slightly flat; it presumes that the crater is glacier filling. Although there is no volcanic activity or hot fumeroles lately, there are reports of SO_2 gas in higher elevations (Hall et al., 2012). Most of the stratigraphy is composed by dark layers of ash and andesite scoria, which is product of the fall of eruptive clouds with intercalations of fluvial deposits (Hall et al., 2012).

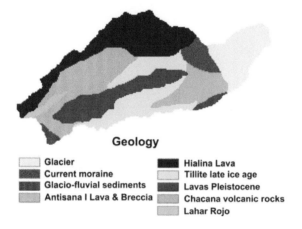

Geology

Glacier	Hialina Lava
Current moraine	Tillite late ice age
Glacio-fluvial sediments	Lavas Pleistocene
Antisana I Lava & Breccia	Chacana volcanic rocks
	Lahar Rojo

Figure 2-7 Geology of the catchment the Los Crespos - Humboldt, Antisana icecap - Ecuador (original source from Hall et al (2012))

Acknowledgements

Special acknowledgment goes to the institutions in Ecuador that cooperated providing key information and data were EPMAPS, EPN, INAMHI, INIGEMM, Ministry of Environment and Herbario Nacional del Ecuador.

We all are very ignorant. What happens is that not all ignore the same things (A. Einstein)

3

CATCHMENT CHARACTERIZATION AND ALTITUDINAL- RANGE ANALYSIS OF CARBON STOCKS

This chapter covers the importance of quantifying carbon stocks in terrestrial ecosystems, which is crucial for determining climate change dynamics. The present regional assessments of carbon stocks in tropical grasslands are sometimes extrapolated to unsampled areas with a high degree of uncertainty and without considering the carbon and nitrogen composition of vegetation and soil along altitudinal ranges. Here we carried out an altitudinal - range analysis of the main ecophysiological parameters that were not readily available in literature for this specific *páramo* ecosystem. It was important to scientifically prove the potential distribution of these parameters along and altitudinal range and confirm that the elevation plays an important role on the distribution and diversity of the main growth forms of vegetation; being tussock, acaulescent rosette, and cushions. In addition, we quantified the above-ground biomass and soil organic carbon stocks and its relationship with soil texture.

This chapter is based on:

Minaya, V., Corzo, G., Romero-Saltos, H., van der Kwast, J., Lantinga, E., Galarraga-Sanchez, R. and Mynett, A.E.: Altitudinal analysis of carbon stocks in the Antisana *páramo*, Ecuadorian Andes, Plant Ecology, 9, 5, 553-563, doi:10.1093/jpe/rtv073, 2015

3.1 Introduction

Approximately 25% of the world's terrestrial ecosystems are grasslands distributed from arid to humid regions (Ojima et al., 1993). Grasslands are biomes that contain large belowground carbon stocks in roots and soil organic matter (Verschot et al., 2006), related to interannual climatic variability and differences in land management practices (Connant et al., 2001). The IPCC 2013 Guidelines (IPCC, 2013) highlight the importance of quantifying above- and belowground carbon stocks and emissions from grasslands due to burning and land-use change. However, such recommendations do not include the influence of local factors such as altitude, vegetation and soil texture on carbon stocks, which are necessary in order to understand the complex biogeochemical interactions between soil and plant community (Osanai et al., 2012). For example, small clay particles favor water-holding capacity and nutrient contents of the soil (Jobbagy and Jackson, 2000) and enhance carbon sequestration by stabilizing soil organic carbon (Paul, 1984).

The largest montane grasslands in the Neotropics occur in high-altitude ecosystems known as *páramos*. These are mostly distributed in the northern Andes (Hofstede et al., 2003; Myers et al., 2000) and represent 9.4% of the global grassland area (White et al., 2000a). The *páramos* contain many microhabitats, hold enormous biodiversity and supply important ecosystem services such as water regulation, water provision and carbon storage (Buytaert et al., 2006a; Myers et al., 2000). Despite their importance, they are subjected to a high rate of degradation due to grazing, burning and land conversion (Hofstede, 1995; Verweij and Budde, 1992).

Many studies have emphasized the environmental controls on the dynamic pool of carbon in high-latitude (tundra) and high-altitude (*páramo*) grasslands as a crucial component of the global terrestrial carbon budget, but the effects of vegetation and soil on carbon dynamics in these ecosystems remain unresolved (Alvarez and Lavado, 1998; Anthelme and Dangles, 2012; Körner, 2007; Parton et al., 1987; Paruelo et al., 1997; Paul, 1984; Tonneijck et al., 2010). In particular, soil carbon storage in *páramos* can be large because they occur on volcanic soils (Andosols) (Tonneijck et al., 2010); indeed, it is probably much greater than that contained in vegetation (Kadovic et al., 2012).

A common practice to estimate carbon stocks in *páramos* usually takes into account only the most common or dominant vegetation: the tussocks from the family Poaceae. However, it is unknown if this methodological approach is accurate enough to estimate the carbon stocks in this ecosystem, given the plethora of species and growth forms that *páramos* contain. Additionally, there is not much information on how the soil – vegetation interaction affects the carbon stocks along altitudinal ranges, which in turn should influence carbon estimates at the landscape level. The present study attempts to overcome these limitations by (i) describing the vegetation and soil characteristics of a *páramo* grassland, (ii) using this information to selectively sample carbon and nitrogen concentrations in the living biomass of the most dominant plant growth forms and (iii) determining carbon and nitrogen concentrations in the topsoil underneath the plants sampled.

3.2 Materials and methods

3.2.1 Sampling design

During the reconnaissance survey prior fieldwork, a difference in the amount and biomass and type of vegetation was noticed not only along an altitudinal gradient (vertical) but also along a transverse axis (horizontal). Sampling was conducted during November and December 2012 in the catchment described in section 2.2. The sampling strategy was based on a digital elevation model and a topographic map 1:25 000 (IGM, 1990) of the Los Crespos Humboldt basin. The idea was not to concentrate only in the dominant factor but it was also to grasp the variability within the altitudinal zone. Sampling followed a random sampling (Daniel, 2012) assuming homogeneity at three altitudinal ranges. A total of 27 permanent sampling plots, each measuring 15 × 15 m were divided into 2 × 2 m subplots (except for the subplots at the corners). The plots were randomly established in the low, mid and high areas of the basin (Figure 3-1). These three sampling areas, here called catchments, roughly correspond to three altitudinal ranges: low catchment (n = 10 plots, 4000–4200 m a.s.l.), mid catchment (n = 10, 4200–4400 m a.s.l.) and high catchment (n = 7, 4400–4600 m a.s.l.) (Figure 3-1).

The plots were established facing north using a compass (SILVA professional S15T DCL) and the corners were marked with PVC tubes. The slope, altitude and geographic coordinates of each plot were measured using a clinometer (CL1002), an altimeter (Multi Function Altimeter with Digital Compass) and a GPS (Magellan eXplorist 100), respectively.

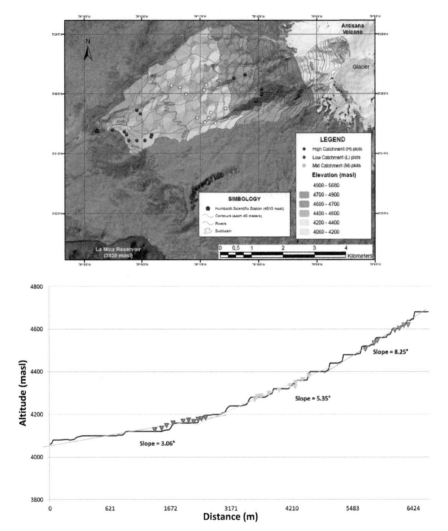

Figure 3-1 a) Sampling plots (*n*= 27) are shown as crosses (red = low catchment plots, green = mid catchment plots, blue = high catchment plots), b) longitudinal profile of the basin.

The floristic inventory was conducted in those plots that did not fall in an area with evident signs of burning or permanent grazing. The sampling method of Mena (1984) was applied to count all individuals per species, and the Braun-Blanquet (1964) method was used to estimate the area covered by all vascular and nonvascular plant species in each 2 × 2-m subplot. For a given area, this method classifies the area covered by a species in seven percentage categories (%): <1% cover, 1–5, 5–15, 15–25, 25–50, 50–75 and ≥75%. Plants were further sorted into growth forms according to the classification for Ecuadorian *páramos* proposed by Ramsay and Oxley (1997): stem rosette, basal rosette, tussock, acaulescent rosette, cushion, upright shrub, prostrate shrub, erect herb, prostrate herb and creeping herb. The cover was determined as the area occupied by the base of the plant. We measured the two horizontal dimensions and applied the formula of an ellipse to calculate the grassland basal area (*GBA*) for all vegetation growth forms. Botanical vouchers were collected for vascular plants only, as deemed necessary for taxonomical identification (Cleef, 1981; Cuatrecasas, 1968; Rangel et al., 1983; Ronquillo., 2010); vouchers were identified at the National Herbarium of Ecuador (QCNE) and the Catholic University's Herbarium (QCA), both in Quito.

Finally, for each of the main growth forms in each plot, we sampled leaves, leaf litter and roots following standard protocols (Law et al., 2008). In each plot, one topsoil sample (0–0.3 m depth) was collected in each cardinal point, 3 m apart from the edge, using a 70-mm soil corer (EIJKELKAMP, model GM085). Samples were collected into clearly labeled dark polyethylene bags and later refrigerated.

3.2.2 Laboratory analyses

3.2.2.1 Carbon and nitrogen analyses

Leaves, litter and roots were dried at 60°C for 48 h. Soil was dried at room temperature. All dry plant and soil samples were grinded in a mill (Retsch Mixer Mill MM200) at a vibrational frequency of 300 min^{-1} for 4 min to obtain a homogeneous sample and then dried again at 60°C overnight. These samples, placed on a tarred tin capsule, were weighed to the nearest hundredth of a gram using a digital lab scale balance (Mettler Toledo AE100). Carbon and nitrogen concentrations in plant and soil samples were measured by an EA1108 CHN-O Element Analyzer (Fisons Instrument) using the Dumas Combustion Method (Dumas 1826; cited by Buckee 1994).

3.2.2.2 Lignin and cellulose in plants

To measure lignin and cellulose in roots and litter, dry samples were grinded, sieved through a 1-mm mesh and then placed in a F57 filter bag. Lignin and cellulose concentrations were determined using acid detergent fiber methods (ADF, 1990) via ANKOM Technology - 01/02 and - 07/02. Lignin (%) and cellulose (%) were calculated as:

$$Lignin\ (\%) = \frac{W_4 - (W_1 C_2)}{W_2} 100 \tag{3-1}$$

$$Cellulose\ (\%) = \frac{W_3 - (W_1 C_1)}{W_2} 100 - Lignin\ (\%) \tag{3-2}$$

where W_1 is the bag tare weight; W_2 is the sample dry weight; W_3 is the weight after the extraction process; W_4 is the weight of organic matter (loss of weight of bag ignited at 525 °C for 3 hours, plus fiber residue); C_1 is the blank bag correction (final oven dried weight/original blank bag weight) and C_2 is the ash corrected blank bag (loss of weight on ignition of bag/original blank bag weight).

3.2.3 Data treatment and analyses

3.2.3.1 Altitudinal effects on growth forms and plant/soil chemical composition

To better understand the chemical composition of plant components responses to altitudinal ranges and growth forms of vegetation in more detail, the distribution of the chemical composition within altitudinal ranges was examined using a multiple comparison test. Boxplots of the different carbon, nitrogen, cellulose and lignin concentrations of the main growth forms of vegetation at each altitudinal range display the distribution of data based on quartiles, maximum and minimum values and outliers. A multiple comparison test was chosen to identify the differences in the chemical composition of the plant and soil components at three altitudinal ranges. Tukey multiple comparisons tests (Tukey, 1991) were used to compare carbon and nitrogen pools among different altitudinal ranges. Cellulose and lignin concentrations along the altitudinal range were also analyzed. In addition, for each growth form, Pearson correlation coefficients were calculated to test for significant relationships among species diversity, soil texture composition and nitrogen and carbon concentrations. Pairwise P values were corrected for multiple inferences using Holm's method (Holm, 1979).

To represent the landscape variation of carbon and nitrogen concentrations in different vegetation growth forms along the altitudinal range, a multidimensional scaling (MDS) ordination method was utilized. Such matrix was based on a Euclidean dissimilarity matrix (Anderson, 2006; Legendre and Legendre, 1998) computed from carbon and nitrogen concentrations (%) in the topsoil (0–0.3 m depth). The dissimilarity matrix was also subject to testing for altitudinal range effects using permutational multivariate analysis of variance (PERMANOVA) (Anderson, 2001). All statistical analyses were done with R 2.12.0 (R Development Core Team., 2007), using the vegan package (Oksanen et al., 2011) and the plotrix package (Lemon, 2006).

3.2.3.2 Aboveground biomass and carbon stocks

Biomass for tussocks, acaulescent rosettes and cushions—the main growth forms occurring in the study area—were calculated by building allometric equations (Figure 3-2). These equations relate GBA (m²), represented by the area of an ellipse (Ganskopp and Rose, 1992), and aboveground biomass (B, kg), represented as the dry weight (Gomez-Diaz et al., 2011). To build the allometric equation for each growth form, weighted linear regression models on log-transformed data were used.

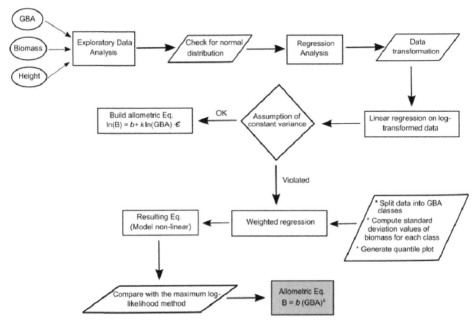

Figure 3-2 Simplified scheme of the statistical steps followed to obtain allometric equations for the main growth forms of *páramo* vegetation starting from data measured in situ.

This model was nonlinear and adjusted for the variation among classes of GBA. GBA and B were related by the following regression equation:

$$B = b\,(GBA)^{\,k} \tag{3-3}$$

where b and k are parameters estimated statistically. Analyses were conducted in R (R Development Core Team., 2007).

In order to calculate the aboveground carbon stocks (kg C/m²), we first estimate the aboveground biomass per square meter and then multiply for the average percentage of carbon in leaves for each growth form at each altitudinal range.

To extrapolate the aboveground carbon stocks for the entire basin, we (i) related the dominant growth form of each sampling plot with its NDVI (normalized difference vegetation index), (ii) used this relationship to determine the rest of the pixels in the basin and (iii) calculated the percentage of coverage of each growth form in the different altitudinal ranges (Table 3-1). The NDVI was calculated from an ASTER (Advanced Spaceborne Thermal Emission and Reflection Radiometer) image (this relationship was feasible because the pixel size was the same as the sampling plot: 15 × 15 m). NDVI was validated with field measurements of radiance (W/sr/m²) taken with a spectroradiometer (ASD FieldSpec 4 HiRes) in all sampling plots.

3.2.3.3 Soil organic carbon stocks

The total soil organic carbon stocks (TCS_d) per square meter (kg C/m²) in a soil sample of depth d (m) was estimated using an equation from a case study in the northern Ecuadorian *páramos* (Tonneijck et al., 2010):

$$TCS_d = \sum_{i=1}^{k} {}^{b}\rho_{s,i}\, C_i\, D_i\, (1 - S_i) \qquad (3\text{-}4)$$

where k is the number of soil layers up to depth d (here, $k = 1$ because we only considered a single 0–0.3 m layer), $\rho_{s,i}$ is the dry bulk density (b) of soil (s) layer i (kg/m3), C_i is the carbon concentration (fraction) in layer i (kg/kg), D_i is the thickness of layer i (m) and S_i is the fraction of soil fragments ≥2 mm in layer i (kg/kg), a term that can be neglected in the case of *páramo* soil (Tonneijck et al., 2010). To estimate soil dry bulk density (${}^{b}\rho_{s,i}$), we used its relationship to carbon fraction in layer i (C_i) as showed by Tonneijck et al. (2010):

$$ {}^{b}\rho_{s,i} = 0.9247\, e^{-5.497 C_i} \qquad (3\text{-}5)$$

The estimation of the total soil organic carbon stocks for the entire basin only considered the dominant growth form of each pixel.

Table 3-1 Dominant growth form coverage (%) in the LCH basin according to a supervised classification of an ASTER image (15 × 15 m pixels).

Pixel classification	% of growth form coverage in different altitudinal ranges (catchments)		
	low	mid	high
Tussocks	59.7	56.6	45.4
Acaulescent Rosettes	28.8	36.9	20.6
Cushions	7.0	3.2	1.1
Other (including minor growth forms, bare soil or rock)	4.5	3.3	32.9

3.3 Results

3.3.1 Altitudinal analysis of carbon and nitrogen pools

3.3.1.1 Soil and litter

Soil carbon concentrations (% C) were significantly higher at the low catchment which is dominated by tussocks and acaulescent rosettes (Figure 3-3). A similar pattern was observed with soil nitrogen concentrations (% N). Cellulose in litter showed higher concentrations at the low catchment for acaulescent rosettes and cushions only. In contrast, for these same growth forms, lignin in litter showed the highest concentration at the high catchment. Tussocks did not show any clear altitudinal pattern of litter cellulose and litter lignin concentrations. For soil under tussocks and acaulescent rosettes, the averaged ratio between soil carbon and soil nitrogen was 16:1, but for the soil under cushions this ratio can go up to 19:1 in the high catchment (Table 3-2).

Table 3-2 Comparison of carbon and nitrogen ratios of the main growth forms of *páramo* vegetation in the Los Crespos Humboldt basin, analyzed per altitudinal ranges or catchments (low, mid and high).

Parameter	Tussocks			Acaulescent Rosettes			Cushions		
	low	mid	high	low	mid	high	low	mid	high
C:N in soil	16:1	16:1	16:1	16:1	16:1	16:1	16:1	14:1	19:1
C:N in leaves	93:1	64:1	72:1	25:1	28:1	24:1	22:1	26:1	26:1
C:N in roots	23:1	27:1	23:1	28:1	25:1	32:1	28:1	25:1	39:1

3.3.1.2 Live tissues (leaves and roots)

Percentage of C in leaves and roots did not show any particular trend, with exception of leaf % C of acaulescent rosettes, which showed the highest values at the low catchment (Figure 3-3). Percentage of N in roots was always lower at the high catchment. Cellulose in roots in general followed this same tendency but not in tussocks. Conversely, lignin in roots tended to have higher concentrations at the high catchment, although this was only significant for tussocks (Figure 3-3).

The averaged C:N ratio in tussock leaves varied from 64:1 in the mid catchment to 93:1 in the low catchment (Table 3-2). The C:N ratios for acaulescent rosette and cushion leaves were similar and varied around 25:1 regardless of altitudinal range.

For all growth forms, the C:N ratio of roots varied around 26:1 in the low and mid catchments, while for the high catchment, the ratio increased to 32:1 for acaulescent rosettes and 39:1 for cushions (Table 3-2).

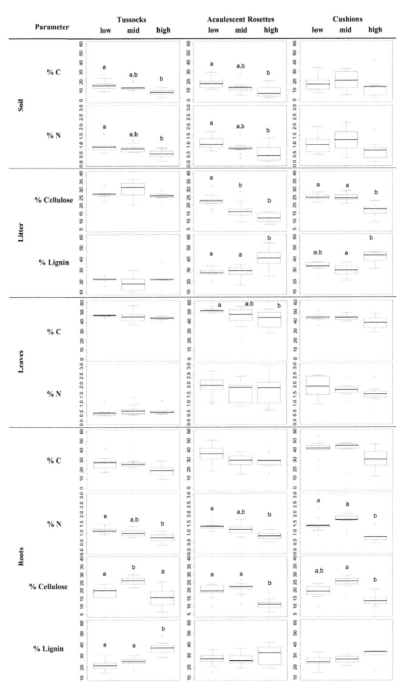

Figure 3-3 Carbon, nitrogen, cellulose and lignin concentrations of the main growth forms of *páramo* vegetation in the Los Crespos Humboldt basin, analyzed per altitudinal catchments (low, mid and high). Lowercase superscripted letters indicate significant differences among altitudinal ranges ($P \leq 0.05$), according to Tukey's test.

3.3.1.3 Relationships to soil texture

Soil texture variables (e.g. % silt and % sand) were highly correlated with altitude ($r_{altitude} - \%\ silt = -0.89$, $P_{adj} < 0.001$; $r_{altitude} - \%\ sand = 0.88$, $P_{adj} < 0.001$). In addition, in acaulescent rosettes but not in tussocks, soil % C and soil % N showed a high correlation with % silt, % sand and altitude (Table 3-3). High soil % C and high % N occurred at low altitudes, where relatively small soil particles were found. Cellulose concentration in litter for acaulescent rosettes and cushions increased with increasing % silt in soils but decreased as sandy soils became more dominant at higher altitudes. This pattern was completely opposite to what was observed with lignin concentration in litter.

In live tissues, root % N for all three growth forms increased with decreasing % sand, which increased at higher altitudes (Table 3-3). The % cellulose in roots was not related to soil texture nor altitude, except for acaulescent rosettes, in which it decreased with increasing altitude. The % lignin in roots for tussocks increased with increasing % sand and altitude.

Table 3-3 Pearson correlation matrix between those parameters that showed statistical significance in Figure 3-3 and variables of altitude and soil texture.

	Growth form	Parameter	% Silt	% Clay	% Sand	Altitude
Soil	TU	% C	0.58	-0.29	-0.59	-0.55
		% N	0.55	-0.26	-0.58	-0.54
	AR	% C	0.59*	-0.45	-0.57*	-0.58*
		% N	0.60**	-0.44	-0.59*	-0.59**
Litter	AR	% Cellulose	0.55*	-0.42	-0.54*	-0.71***
		% Lignin	-0.66**	0.41	0.67***	0.62**
	CU	% Cellulose	0.76**	-0.32	-0.79***	-0.78***
		% Lignin	-0.77**	0.28	0.81***	0.64
Leaf	AR	% C	0.55*	-0.51	-0.51	-0.51
Root	TU	% N	0.64	-0.25	-0.68*	-0.70*
		% Cellulose	0.49	-0.11	-0.55	-0.36
		% Lignin	-0.76**	0.33	0.80***	0.86***
	AR	% N	0.63**	-0.46	-0.62**	-0.64**
		% Cellulose	0.62**	-0.38	-0.64**	-0.57*
	CU	% N	0.75**	-0.49	-0.74**	-0.64
		% Cellulose	0.54	-0.29	-0.55	-0.44

Abbreviations: AR = acaulescent rosettes; CU = cushions; TU = tussocks.
*$0.01 \leq P \leq 0.05$, **$0.001 \leq P \leq 0.01$, ***$P \leq 0.001$.

3.3.2 Aboveground biomass and carbon stocks

The following allometric equations for aboveground biomass (B, kg) estimates of the main growth forms of vegetation, based on GBA (m²), were developed:

For tussocks ($n = 46$ plants): $B = 3.89 \times 10^{-1} \times GBA^{0.158145}$ (3-6)

For acaulescent rosettes ($n = 74$): $B = 4.51 \times 10^{-2} \times GBA^{0.28599}$ (3-7)

For cushions ($n = 29$): $B = 7.56 \times 10^{-3} \times GBA^{0.703931}$ (3-8)

The aboveground biomass was higher at the low catchment for tussocks (3.7 kg/m²) and cushions (1.2 kg/m²). Acaulescent rosettes did not show a clear altitudinal pattern for aboveground biomass (Table 3-4).

Table 3-4 Total aboveground soil carbon stocks, total soil organic carbon stocks (0–0.3 m) and total aboveground biomass for the main growth forms at different catchment altitudes (low, mid and high) in the Los Crespos Humboldt basin.

	Total aboveground carbon stocks (kg C)				Total soil organic carbon stocks (kg C)				Total aboveground biomass (kg m²)			
	low	mid	high	Total	low	mid	high	Total	low	mid	high	Total
Tussocks	3.3×10^6	1.9×10^6	2.2×10^5	5.4×10^6	3.3×10^7	4.2×10^7	1.8×10^7	9.3×10^7	3.7	1.7	0.44	5.8
Acaulescent Rosettes	2.7×10^5	5.7×10^5	1.8×10^5	1.0×10^6	1.6×10^7	2.6×10^7	7.6×10^6	5.0×10^7	0.6	0.8	0.6	2.0
Cushions	1.2×10^5	4.6×10^4	1.0×10^3	1.7×10^5	3.7×10^6	2.1×10^6	3.6×10^5	6.2×10^6	1.2	0.8	0.1	2.1
Total	3.7×10^6	2.5×10^6	4.0×10^5	6.6×10^6	5.3×10^7	7.0×10^7	2.6×10^7	1.5×10^8	5.5	3.3	1.1	9.9

Area-based aboveground carbon stock in tussocks (2.86 kgC/m²) was significantly higher at the low catchment than at higher altitudes (2.43–2.37 kg C/m²) (Figure 3-4). Acaulescent rosettes and cushions contained similar area-based aboveground carbon stocks in the entire basin (1.05–1.33 kg C/m²). Also, the total aboveground carbon contained in tussocks was 5 times greater than that contained in acaulescent rosettes and 32 times greater than that contained in cushions. The total aboveground carbon stock for all three growth forms in the basin was estimated as 6.6×10^6 kg C (Table 3-4).

3.3.3 Soil organic carbon stocks

Area-based soil organic carbon stocks (kg C/m²) were significantly higher in the low and mid catchments for all growth forms (Figure 3-4). Total soil carbon stocks in tussocks were 2 times greater than acaulescent rosettes and 15 times greater than cushions for the entire basin. The total soil organic carbon stock in the basin was estimated as 1.5×10^8 kg C (Table 3-4).

3.3.4 MDS analysis

The MDS revealed a similarity of sites located at the same altitudinal range (color coding) (PERMANOVA: $F = 6.73$, df1 = 2, df2 = 76, $P < 0.001$) (Figure 3-5). MDS showed that higher values of % C in roots were restricted to acaulescent rosettes and cushions. The % cover along with the area-based aboveground carbon stocks for tussocks increased in the low catchment. Sandy soils increased along an altitudinal range, whereas the greatest amount of soil organic carbon stocks was available at the low catchment.

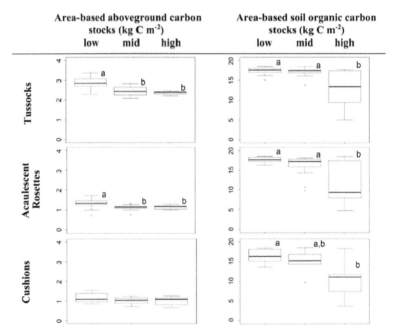

Figure 3-4 Comparison between the area-based aboveground carbon stocks and area-based soil organic carbon stocks for the main growth forms at different catchment altitudes (low, mid and high) in LCH basin. Lowercase superscripted letters indicate significant differences among altitudinal ranges ($P \leq 0.05$), according to Tukey's test.

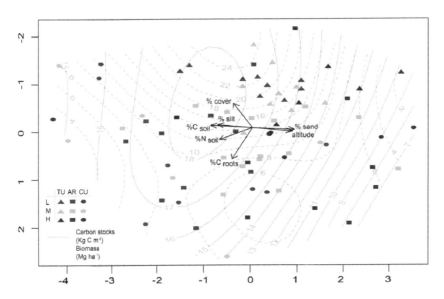

Figure 3-5 MDS of carbon composition (total stress 14.49%). MDS translates Euclidean dissimilarity computed from normalized carbon stocks into distance among sites and growth forms of vegetation into a two-dimensional plot with equally weighted dimensions. Sites are coded according to growth forms (TU = tussocks, AR = acaulescent rosettes, CU = cushions) and altitudinal ranges (L = low catchment, M = mid catchment, H = high catchment) as colors. Arrows indicate the strength and direction of the gradient of the variables indicated.

3.4 Discussion

3.4.1 Altitudinal analysis of carbon and nitrogen pools

3.4.1.1 Soil and litter

The soil carbon and soil nitrogen concentrations in cushions do not show a clear altitudinal pattern. This might be not only because they are found in many habitats, from rock formations to swampy areas, but also because they obtain water and nutrients from litter decomposition and peaty mass trapped within their own mounded formations (Kleier and Rundel, 2004; Sklenár, 1998). On the other hand, soil carbon and soil nitrogen concentrations in tussocks and acaulescent rosettes are higher at low elevations. Their homogeneous growth enhances soil microbial activity and nutrient conservation, in contrast with soils at high catchments that are less vegetated and thus have higher nutrient losses (O'Connor et al., 1999).

The cellulose and lignin found in litter are crucial for the regulation of litter decomposition both in the early and later stages (Fioretto et al., 2005; Rahman et al., 2013). They are also important for soil organic matter by providing energy input for soil microflora and fauna (Bernhard-Reversat and Loumeto, 2002; Fiala, 1979) and even for local atmospheric composition through the emission of CO_2 by microbial activity (Chen, 2014; Vanderbilt et al., 2008). This physicochemical quality and composition of plant litter have strong effects on ecosystem functioning. Further, high C:N ratios enrich soil organic carbon, which increases the amount of carbon in the soil, therefore playing an important role in the global carbon cycle (Bell and Worrall, 2009).

3.4.1.2 Live tissues (leaves and roots)

The high concentration of carbon in leaves for acaulescent rosettes makes them decompose at a rapid rate. This drives the nutrient turnover and controls the micro-atmospheric composition of the plant at low catchments.

Although the carbon measured in tussock roots was low on average, tussocks have roots around 1.5 m long, where carbon can be stored. These lengthy tussock roots usually help to stabilize slopes, control erosion and increase soil porosity for water absorption. On the contrary, acaulescent rosettes have thick taproots ~0.15 m long, while cushions have roots around 1 m long, depending on the species and soil texture (Attenborough, 1995).

The cellulose found in roots was very similar among the three growth forms of vegetation, giving the same strength to their leaves and rigidity to the cells and making them difficult to break down. Cellulose gives the physical support required not only to anchor the plant but also to absorb water and minerals. Lignin also strengthens the cell wall and plays an essential role not only for water transport within the plant but also in the carbon cycle because it sequesters atmospheric carbon in living tissues for relatively long periods. Lignin found in tussock roots increased with altitude ($P < 0.001$), while the same trend was observed for acaulescent rosettes and cushions. The high lignin concentration in the species located in the high catchment could explain the survival of certain species to such relatively hostile environment, filled with loose stones, sandy soil, low soil water-holding capacity and low nutrient content (Körner, 2003; Molau, 2004; Rada et al., 2001). Additionally, they have to cope with climatic conditions of low oxygen concentration, low rainfall, high solar radiation, strong wind abrasion and night frost (Baruch, 1984).

In general, tussocks leaves have slower decomposition rates than acaulescent rosettes and cushions, at low and mid catchments. This can be attributed to their high C:N ratios and, at the high catchment, to high lignin concentration. Acaulescent rosettes and cushions leaves show a constant C:N ratio of 25:1 throughout the altitudinal range, which should promote rapid decomposition by microbial activity.

3.4.1.3 Relationships to soil texture

Most soils in the Ecuadorian highlands are derived from volcanic material (Hall and Beate, 1991); however, soil particle sizes differ among altitudinal ranges. In this study, sandy soils were predominant at higher elevations (>4500 m a.s.l.) (Figure 3-5). This soil drains well but also dries out and warms up rapidly, reducing the capability of holding moisture and providing poor conditions to store plant nutrients (Buytaert et al., 2005c) (Table 3-3). On the other hand, silty soils, mostly located at lower altitudinal ranges, are poor in aeration and percolation but offer high water-holding capacity, which in this study resulted in enriched soil carbon and nitrogen concentrations (Table 3-3).

3.4.2 Aboveground biomass and carbon stocks

Tussocks, acaulescent rosettes and cushions were the dominant growth forms in the study area (as they are in the Ecuadorian *páramos* in general; personal observation). This implies a strong relationship between growth form adaptation and environmental variables, such as high-elevation atmosphere, high radiation and drying wind effects (Ramsay and Oxley, 1997). Several studies (Hedberg, 1992; Ramsay and Oxley, 1997) agree that tussocks and acaulescent rosettes are responsible, on average, for >50% vegetation cover from 4000 to 4400 m a.s.l., which in this study corresponds to the low and mid catchments. Vegetation cover diminishes with altitude because of the less favorable climatic conditions, under which only a few plants can survive and adapt such as those that developed pubescence in order to expand its boundary layer (Sklenar and Ramsay, 2001; Stern and Guerrero, 1997). This agrees with expert local knowledge during our sampling campaign.

Aboveground biomass in tussocks was higher along the low and mid catchments (4000–4400 m a.s.l.) (Figure 3-5). At these altitudes, tussocks benefit from the air that is trapped between the leaves and cool down slowly (Ramsay, 2001). At higher altitudes, tussock biomass is less abundant but becomes important as shelter to other growth forms, enabling them to increase their size (Ramsay and Oxley, 1997). The biomass of acaulescent rosettes and cushions did not show a particular trend with altitude: both grow at the soil surface and remain slightly warmer than night air temperature at night (Hedberg and Hedberg, 1979). Hard and soft-mat cushions maintain the internal temperature stable and tend to be conspicuous in flooding zones or where water is retained (Bosman et al., 1993; Cleef, 1981; Jørgensen and Ulloa, 1994).

It appears that under natural conditions, the aboveground herbaceous biomass for the three growth forms of *páramo* vegetation in the area will not decline below its current abundance and distribution as long as they are not exposed to practices of burning, grazing or any anthropogenic activity (Hofstede, 1995; Laegaard, 1992; Ramsay, 1992; Verweij, 1995). The *páramo* vegetation is resilient to high daily variability in the climatic variables and most probably, the vegetation will continue to capture more carbon in all plant components.

3.4.3 Soil organic carbon stocks

Higher levels of soil organic carbon stocks are found in the low and mid catchments, which could be associated to higher content of soil silt fraction. Silt and clay stabilize the soil organic carbon by protecting organic matter from microbial decomposers (Paul, 1984; Six et al., 2002). The association between soil organic carbon stocks and the change in soil texture with depth is still vague and needs further analysis. It is important to realize, for example, that other regional studies of grasslands (Jobbagy and Jackson, 2000; Spain et al., 1983) also based their results of soil organic carbon concentrations at the top 30 cm of soil. The carbon concentration patterns at higher depths are basically unknown. The increase of carbon stocks in the low and mid catchments (Figure 3-5) was expected because of the predominance of the silt fraction in the soil texture at these elevations (Sacramento et al., 2014).

Note also that the global averaged values for Andosols in tropical grasslands (Jobbagy and Jackson, 2000; Schlesinger, 1986) underestimated the values of soil organic carbon stocks compared to studies conducted *in situ* (Eswaran et al., 1993; Poulenard et al., 2003; Tonneijck et al., 2010), including this one.

3.5 Conclusion

The established relationships among soil, vegetation and altitude shown in this study must be taken into account to estimate both aboveground and soil organic carbon stocks in *páramo* regions. Such estimates will be considerably inaccurate if these relationships are ignored.

1) The high soil nutrient availability in tussocks and acaulescent rosettes at low catchments enhance litter quality and therefore microbial activity. This

physicochemical quality and composition of plant litter have strong effects on ecosystem functioning. In addition, the high C:N ratios enriches the soil organic carbon and therefore playing an important role in the global carbon cycle.

2) Tussocks and acaulescent rosettes are responsible for > 50% of vegetation cover from 4000 to 4400 m a.s.l. This aboveground biomass in the area will not decline below its current abundance and distribution as long as they are not exposed to practices of burning and grazing.

3) Fine textured soils enhanced the accumulation of soil carbon stocks. This emphasizes the importance of the *páramo* vegetation for carbon sequestration as well as other ecological processes.

Acknowledgements

I appreciate the help of Alexandra Toapanta, Carolina Andrade, Francis Minaya, Cathy Rodriguez, Juan Fernando Murriagui, Rocío Sangucho, Julián Aguirre, Christian Gomez, Mercedes Villacis, Marcos Villacis and Hennie Halm, who assisted during fieldwork and laboratory analysis, and Kenny Griffin who proofread the manuscript.

Assumptions imply uncertainty, not desirable but unavoidable
(Anonymous)

4

ESTIMATING GROSS PRIMARY PRODUCTION AND HYDROLOGICAL PROCESSES IN PÁRAMO GRASSLANDS

Despite being essential ecosystems that sustain important ecological processes, just a few efforts have been made to estimate the gross primary production (GPP) and the hydrological budgets along an altitudinal gradient for grasslands in the Andean Region. In this thesis an effort was made to choose an appropriate model that not only has the capability to simulate vegetation processes and fluxes but that also has been widely used in similar case studies at a regional scale, has sufficient theory description on governing equations, and is free software and open source. This chapter applies one of the ecophysiological and biogeochemical models that were found suitable for estimating carbon, nitrogen and water fluxes in these grasslands ecosystems taking into account the main properties of the *páramo* vegetation in plant functional types, site/soil parameters and daily meteorology. The model chosen was the BIOME-BGC that simulates the GPP and the water fluxes in a representative area of the Ecuadorian Andean *páramos*. It focuses on three main growth forms of vegetation and is also extended to cells with similar properties. An accurate estimation of the temporal changes of carbon and water budgets can potentially assess the effect of climate drivers in the biomass productivity of this terrestrial ecosystem.

This chapter is based on:

Minaya, V., Corzo, G., van der Kwast, J., and Mynett, A. E.: Simulating gross primary production and stand hydrological processes of *páramo* grasslands in the Ecuadorian Andean Region using BIOME-BGC model, Journal of Soil Science, 181, 7, 335-346, 2016.

4.1 Introduction

Globally, biogeochemical modelling studies have been increasing in order to accurately quantify carbon, water and energy dynamics and fluxes for terrestrial ecosystems (Schimel et al., 1996; Waring and Running, 2007). These process-based numerical models deal with a large range of parameters and site-specific features, that can increase complexity and difficulty during calibration and validation (Wang et al., 2009). Additionally, there is limited availability of plant-specific information, which requires the ecophysiological models to use a generalized parameterization (Minaya et al., 2015b). This has raised concern about the potentially high uncertainty of the modelled, terrestrial ecosystems' regional responses to spatially heterogeneous vegetation and climate variation. Small-scale studies of biological processes could possibly bridge the gap between organismal, plant-stand and regional understanding of the carbon, water and nitrogen budget (Thornton et al., 2002). In addition, several studies highlight the importance of field measurements to complement the simulation modelling to assess the ecosystem dynamics in an appropriate manner (Bond-Lamberty et al., 2005). However, field measurements of the primary production, outflows, storage pools, ecosystem production or exchange can be difficult, damaging, costly and time-consuming (Clark et al., 2001; Randerson et al., 2002) particularly for mountainous terrains (Dufour et al., 2006). In this regard, information required for high-altitudinal tropical grasslands is not always readily available (FAO, 2010).

The Andes are characterized by high-altitudinal ecosystems better known as *páramos*. The *páramos* are highly biodiverse and supply important ecosystem services such as water regulation, provision and carbon storage (Buytaert et al., 2006a; Myers et al., 2000). They have also been considered as crucial components of the global terrestrial carbon budget. However, the effects of vegetation and soil on carbon and water dynamics in these particular ecosystems are still unclear (Anthelme and Dangles, 2012; Tonneijck et al., 2010).

Many process-based vegetation models currently available but, some may not properly represent the carbon, nitrogen, water fluxes and mass for the vegetation and soil components. Yet, these components are exceptionally important in the high-altitude tropical grasslands. Looking briefly at a few models we see their limitations in this regard. For example, WOFOST (WOrld FOod STudies) has limitations related to the simulation of cold and heat stresses, damage from excess water, hail, strong winds and other extreme conditions (Boogaard et al., 1998). LANDIS-II (Forest Landscape Simulation Model) is intended for broad-scale simulations and detailed representation of nutrient cycling and other ecophysiological processes are not yet included (Aber, 1997; Ollinger et al., 2002). DGVMs (Dynamic Global Vegetation Models) (e.g. HYBRID, IBIS, LPJ-DGVM) were developed to cope with the global problems and are able to show multiple interactions of biosphere-hydrosphere vulnerability to climate and land use change over large domains (Gerten et al., 2004; Sitch et al., 2003). For this model, limitations including generalizing plants to a few functional types, inability to route water among simulated units, clear representation of carbon and nitrogen cycles as well as soil processes still exist (Quillet et al., 2010).

BIOME-BGC is another biogeochemical and ecophysiological model capable of representing highly complex ecological and biophysical process in ecosystems (Thornton et al., 2005; White et al., 2000a). The BIOME-BGC has no intrinsic spatial scale; it mostly depends on the availability of driving variables and fractional vegetation cover. It was developed to estimate storages and fluxes of energy, water, carbon and nitrogen for the vegetation and soil components of terrestrial ecosystems at a daily time scale. It defines the equilibrium condition and the initial state variables of the system in the context of local eco-climatic characteristics and has been widely applied to estimate water and nutrient cycling in forested and non-forested ecosystems (Running and Hunt, 1993; Trusilova and Churkina, 2008). Since, most of terrestrial ecosystems are far from equilibrium conditions due to frequent disturbances like fires, cattle grazing, change of land use (Odum, 1971), these processes cannot be directly simulated in the model. Hence, BIOME-BGC assumes an ecosystem equilibrium condition. Various studies indicated the model has a tendency to overestimate production activity and carbon stocks in long-term simulations because of generalizations in the ecosystem process model and its corresponding parameters (Pietsch et al., 2005; Running and Hunt, 1993). Despite these limitations, BIOME-BGC has been successfully applied from regional (Bond-Lamberty et al., 2009; Kimball et al., 2000; Running and Gower, 1991) to global scales and across temperate (Running & Hunt, 1993) and arctic ecosystems including Canada and North Europe (Engstrom and Hope, 2011; Kimball et al., 2009; Ueyama et al., 2009). Additionally, the model has been tested widely ranging from forest (Churkina et al., 2010; Pietsch et al., 2005; Tatarinov and Cienciala, 2009; Tupek et al., 2010; Turner, 2007) to herbaceous ecosystems (Di Vittorio et al., 2010; Hidy et al., 2012) due to its strong ability to simulate carbon dynamics and stand hydrological processes in ecosystems. A more detailed description of the aforementioned vegetation models are explained in the Methods section.

Given the particular strengths of BIOME-BGC, this study will use it to simulate the *páramo* ecosystem processes; identify variations in gross primary production; and its relationship with the hydrologic balance. This study then provides a detailed account of the model structure and an estimation of water and carbon fluxes along an altitudinal gradient.

4.2 Methods and data

4.2.1 Review of suitable environmental models for alpine grasslands

In the last decade, the development of hydrological and environmental models has increased exponentially. There are several environmental models currently available to represent the most important ecosystem processes with different spatial and temporal scales and with a certain complexity degree of data requirements. However, we describe just a few models based on i) their capability to simulate physical and biological processes (carbon, nitrogen and water), ii) if they are free and open-source software, iii) if they have been used in regional/local scales or similar case studies, and iv) their flexibility to couple with other models. The model was chosen based on the highest score of several criteria detailed in Table 4-1.

BIOME-BGC is an ecosystem process-based model which estimates fluxes, storage of energy (carbon, nitrogen and water) and mass for the vegetation and soil components (Hunt et al., 1996; Running and Hunt, 1993) in a daily and annual time steps. Moreover, the algorithms comprise physical and biological processes that include snow accumulation and melting, plant mortality and fire among others. The daily time step simulates photosynthesis, autotrophic and heterotrophic respiration and hydrological budget. After the flux estimation for one-day period, BIOME-BGC updates the mass stored in the vegetation, litter and soil components. The annual time step assigns the nutrients (carbon and nitrogen) among the leaves, roots, litter and soil (Jensen and Bourgeron, 2001). The weather is the most important factor on vegetation processes and fluxes are estimated based on daily weather conditions, and it requires a set of 34 constants describing each plant functional type (White et al., 2000b). The near-surface daily meteorological parameters required by BIOME-BGC are: maximum, minimum and daily average temperature (°C), total precipitation (cm), daylight average partial pressure of water vapor (Pa), daylight average shortwave radiant flux density (W/m^2) and daylength (s). However, in many cases the only data available are temperature and precipitation. The University of Montana, thorough the NTSG (Numerical Terradynamic Simulation Group) has developed algorithms using MT-CLIM and Daymet that can estimate the radiation and humidity parameters when missing. The MT-CLIM is a computer program that estimates temperature, precipitation, radiation and humidity of one site based on the temperature and precipitation of near-by stations. The closer the stations are the better results will be obtained. The estimation of radiation and humidity are more complex. However, the information of latitude, elevation, slope and site features can be used to estimate the daily total radiation with an error of +/- 15%. The expanded version of MT-CLIM logic, called Daymet uses observations from several near-by stations to estimate the meteorological conditions at the site. BIOME-BGC is used across a wide range from regional to global spatial scales to predict changes in potential vegetation and its advantage is the possibility to link with a hydrological model in order to simulate and predict changes in potential vegetation and water resources under specific climate conditions (Jensen and Bourgeron, 2001). Adaptation of BIOME-BGC is feasible by doing model changes as demonstrated in other studies (Korol et al., 1996; Milesi et al., 2005; Tatarinov and Cienciala, 2006). The model has been successfully applied in the estimation of global net terrestrial carbon exchange and atmospheric CO_2 concentrations (Hunt et al., 1996).

DHSVM (Distributed Hydrology Soil Vegetation Model) was developed around 1990's at the University of Washington and currently is in the new released version 3.0. DHSVM ((Doten et al., 2006) is a distributed hydrological model that represents the effects of topography and vegetation on the water fluxes at a regional scale (Wigmosta et al., 2002; Wigmosta et al., 1994). The model includes two vegetation layers (forest canopy overstorey and understorey) and multiple soil layers (Beckers et al., 2009b). DHSVM has been ranked as high-complexity category compared with other nine similar models due to the model functionality for Forest management (Beckers et al., 2009a). The model does not directly simulate the forest growth but it has the flexibility to couple with other models as input (Beckers et al., 2009a). For the application in site-specific case studies, tailor-made model approaches are necessary (Savenije, 2009). The model has been applied in different regime such as rain, snow and mixed climate conditions at small to medium spatial resolutions at sub-daily timescales for several year simulations. The smallest temporal scale is hourly and it has been used mainly in mountainous catchments (Bowling et al., 2000; La Marche and Lettenmaier, 2001; Storck et al., 1998) and in other sites of complex terrain (Cuo et al., 2008). DHSVM requires intensive information being the most important the Digital Elevation Model (DEM) of the basin, soil textural, vegetation and hydraulic information, meteorological conditions at a sub-daily timestep (e.g. precipitation, temperature, humidity, wind speed & radiation) and other information regarding stream network. Many input files are created using GIS (e.g. soil, vegetation, meteorology, stream files). DHSVM gives a wide range of model outputs including evapotranspiration, water balance, soil moisture, infiltration among others. However, it does not simulate the groundwater flow and does not give any outputs regarding nutrient fluxes and water quality conditions (Beckers et al., 2009a). As Beckers et al. (2009a) explained the main advantage of DHSVM is the great range of catchment hydrology applications. Conversely, the model is difficult to use and it might require a high model parameterization and considerable calibration depending on the level of modeling detail (Beckers and Alila, 2004; Thyer et al., 2004). On top of this, another disadvantage could be the underestimation of low flows due to the deficient simulation of groundwater baseflow and storage as well as in high flows due to the lack of glacial melt estimation (Beckers et al., 2009b). DHSVM is freely available and its open source code.

LPJ-DGVM is a generalized processed-based vegetation model that can be used from patch to global scale. LPJ (Lund - Postdam - Jena) dynamic global vegetation model (DGVM) was developed to cope with the global problems; it has a broad range of potential applications. LPJ is able to show multiple interactions of biosphere-hydrosphere vulnerability to climate and land use change over large domains (Gerten et al., 2004). The model can simulate the terrestrial vegetation dynamics and the land-atmosphere carbon and water exchanges in a daily time step. In LPJ-DGVM each average plant individual represents the entire regional population of a plant functional type (PFT), the advantage of this model is that the vegetation structure and dynamics might be represented in more detail. However, Gerten et al. (2004) stated that some of the PFT parameters had to be modified so they can simulate more accurate. Gross primary production is calculated based on the photosynthesis - water balance coupling. Gerten et al. (2004) proved that LPJ output of runoff and evapotranspiration, including the extension of the biases are comparable with the ones calculated by other global hydrological models. As they basically depend on the uncertainties of the climate input data. However, in some regions runoff can be under or overestimated in comparison to the observations (Gerten et al., 2004). Smith et al. (2001) developed a landscape model version called

LPJ-GUESS with the same land-atmosphere coupling in order to assess the impact of climate change and management decisions on vegetation, ecosystems, biodiversity and biogeochemistry for local to regional applications. The model requires climate, atmospheric conditions, soils features and vegetation species traits as input data (Hickler et al., 2004).The advantage of LPJ-GUESS is that it does not require site-specific calibration; therefore it is viable to project vegetation dynamics under different environmental conditions at other sites without modifying the internal parameterization (Hickler et al., 2004). Arneth et al. (Arneth et al., 2007a; Arneth et al., 2007b) highlights the unique feature of the LPJ-GUESS model, that through algorithms takes into account the effect of BVOC (biogenic volatile organic compounds) emitted by vegetation and linked with photosynthesis that no other land surface model has considered so far. LPJ models have adopted bioclimatic limits from the BIOME model versions (Haxeltine and Prentice, 1996; Kaplan, 2001) and it has been applied successfully in a wide range of climate-driven simulations (Hickler et al., 2004; Koca et al., 2006; Miller et al., 2008; Schroter and et al., 2005; Smith et al., 2008; Wolf et al., 2008). Nowadays, the LPJ-DGVM source code and information is freely available but there is no source code open access for LPJ-GUESS. Daily precipitation can be obtained by using a stochastic weather generator. LPJ is being adapted as a component for other climate-carbon cycle models (Joos et al., 2001; Prentice et al., 2001).

WOFOST is a one-dimensional dynamic simulation model of potential and water-limited production situations for a quantitative analysis of a size-limited region. This program was created by the Center for World Food Studies (CWFS) in cooperation with the University of Wageningen and the Center of Agribiological Research and Soil Fertility but nowadays, it is further developed and maintained by Alterra, University of Wageningen and the Joint Research Center in Italy Agri4Cast (WOFOST 7.1; Boogaard et al. 1998). WOFOST analyzes crop growth based on processes such as photosynthesis, respiration and the implications in the change of environmental conditions (Boogaard et al., 2011) with a temporal resolution of one day. It was originally developed for the assessment of yield potential in several crops in tropical countries (van Keulen and van Diepen, 1990); nowadays is successfully applied elsewhere. WOFOST is able to calculate biomass for a determined soil type, crop type and weather conditions, however, it causes non-linear responses that can be partially solved by splitting the spatial domain into small spatial units and the inputs of soil, crop and weather are assumed constant. WOFOST contains three sub-models that compute the moisture content in the root zone of the water balance for different situations, one used for the potential production, for water-limited production and the last one for nutrient-limited production. Based on experience, WOFOST must be carefully used to ensure correct results, and consequently the processes involved should be performed individually (integration, driving variables and rate calculations) (de Wit, 1993; Passioura, 1996). However, WOFOST aims to integrate knowledge on plant growth processes and to test hypothesis by mathematical reproduction. It must be taken into account that WOFOST has some limitations, and it cannot simulate cold and heat stresses, damage from excess water, damage by hail and strong winds and other extreme conditions that might exist (Boogaard et al., 1998). WOFOST has been used in several research studies and published more than 30 journal papers related to climate change, regional yield forecasting, crop yield analysis, model intercomparison among others. PyWOFOST is an implementation of WOFOST, which uses kernel routines that are compiled and linked with python interpreter. So that, PyWOFOST enables the link between the simulations model WOFOST and satellite observations. The importance of PyWOFOST lies in its ability to communicate with databases and its flexibility of adaptation to specific situation.

RHESSys (Regional Hydro-Ecologic Simulation System) is a physically-based hydrologic routing process and GIS-based model that simulates the spatio-temporal interactions of carbon, water and nutrient fluxes on a catchment scale (Band et al., 1993; Jensen and Bourgeron, 2001). The model has been adapted from a previous version of ecophysiological models such as BIOME-BGC and an explicit routine model from DHSVM (Wigmosta et al., 1994) to model saturated subsurface. The spatial data required are used to create the flow paths and to represent the landscape in a hierarchical structure including: basins (stream routing), hillslopes (lateral soil moisture fluxes), zones (climate processing), patches (vertical soil moisture fluxes; soil nutrient processing) and canopy strata (aboveground vegetation carbon, nitrogen, water fluxes). Time series data required are daily precipitation, minimum and maximum temperature, daily streamflow, among others.

Table 4-1 Evaluation of the environmental models considered.

No.	Criteria	BIOME-BGC	WOFOST	LPJ-DGVM	RHESSYS	DHSVM
1	Ample information (user's manual, guides, tutorials, supporting web)	2	2	2	2	2
2	Capability to simulate vegetation processes & fluxes (carbon, nitrogen, water)	2	2	2	2	0
3	Capability to simulate regional climatic processes	1	1	1	1	1
4	Free software	2	0	0	2	2
5	Open source	2	2	2	2	2
6	*páramo* processes (possible to include in the model?), precipitation , etc	1	1	1	1	1
7	Source code (easy to read, self explanatory)	1	2	1	1	1
8	Model used in regional/local scale	2	2	2	0	2
9	Model used in similar case studies	1	1	1	0	0
10	Represent the most important ecosystem processes in a computationally efficient framework	2	1	1	1	2
11	Accurancy in simulations	1	1	1	1	2
12	Capability of model other disturbances	2	0	1	2	1
13	Adequate temporal resolution	2	1	2	2	2
14	Data required to describe plant functional types	2	2	2	1	2
	total points	23	18	19	18	20

Scores for the models: 0 = *does not satisfy*, 1 = *partly satisfies*, 2 = *fully satisfies*

4.2.2 Data availability

As demonstrated in chapter 3 (Minaya et al., 2015a), there are significant differences in the potential distribution of carbon, nitrogen in soil, litter and live tissues concentrations in this region. These were classified into three different altitudinal ranges and verified these findings using other studies at different stratums in the Andean highlands (Poulenard et al., 2001; Sklenar and Jørgensen, 1999). Thus, the biogeochemical BIOME-BGC model is set up according to three altitudinal ranges based on the three dominant growth forms of vegetation across the *páramo* ecosystem. The three altitudinal ranges are identified as Low (4000-4200 m a.s.l.), Mid (4200-4400 m a.s.l.), and High (4400-4600 m a.s.l.) (Figure 3-1) and Figure 4-1 shows the distribution of the growth forms of vegetation throughout the basin.

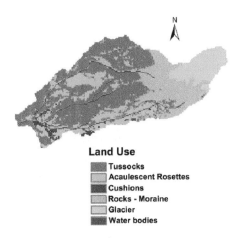

Figure 4-1 Land use distribution of the Los Crespos - Humboldt basin in Ecuador.

The soil in the *páramos*, known as andosols, are developed from volcanic material characterized by high soil moisture, elevated organic C and mineralogical composition (Buytaert et al., 2005c; Buytaert et al., 2005d). The site and soil characteristics for the study area at three altitudinal elevations are specified in Table 4-2.

Table 4-2 Site and soil characteristics (Annual means ± Standard deviation). Meteorological information based on a 12-year daily data.

Parameter	Low (4000-4200 masl)	Mid (4200-4400 masl)	High (4400-4600 masl)
	Site and soil		
Site latitude (°)	-0.4665	-0.4665	-0.4665
Albedo (DIM)	0.1723	0.1759	0.1753
Effective soil depth (m)	1.7	1.0	0.5
Sand:silt:clay ratio	19:66:15	24:59:17	58:20:22
	Meteorological data		
Mean annual air temperature (°C)	7.31±1.44	6.53±0.35	4.82±0.37
Mean annual precipitation (mm)	925.1±100.8	1337.4±196.0	1176.2±184.7

Abbreviations: IRD = Institut de recherche pour le développement, INAMHI = Instituto Nacional de Meteorología e Hidrología - Ecuador

Daily total precipitation and daily maximum and minimum temperatures for a 12-year period (2000-2011) (Table 4-2) were derived from daily meteorological data (Refer to section 2.5). Using data collected from two stations, one at 4000 m a.s.l.

(outlet of the basin) and the other at 4785 m a.s.l. (upper basin) (Figure 2-6) and a mountain climate simulator MT-CLIMB version 4.3 (Kimball et al., 1997; Thornton et al., 2000; Thornton and Running, 1999), short wave radiation and vapour pressure deficit were derived. The mountain climate simulator (MTCLIM) is a model for non-linear extrapolation of meteorological data in high mountains (Hungerford et al., 1989; Running et al., 1987).

MTCLIM requires of initial site-specific information such as: latitude, elevation, slope, aspect, E-W horizon angles, maximum and minimum temperature lapse rate and climate information at a daily basis containing yearday, maximum and minimum temperature, dew point and precipitation. The model uses a point of measurement as a "base" station and the study site of interest taking into account the corrections for the differences in elevation, slope and aspect between both stations. MT-CLIMB extrapolated the variables from the stations to the three altitudinal ranges and performed corrections for differences in elevation, slope, and aspect between the station and the new site. Vapour pressure deficit is calculated from the air temperature and the dew point, which is assumed to be equal to the minimum daily temperature. Daily incoming solar radiation is estimated in two steps: 1) calculation of the potential solar radiation based on solar geometric algorithms (Gamier and Ohmura, 1968; Swift, 1976) that allow estimation of beam potential radiation to a surface; 2) application of the principles based on the algorithm of Bristow and Campbell (1984) that relates diurnal air temperature amplitude to atmospheric transmittance. Figure 4-2 shows the MTCLIM subroutines to estimate daily climatic drivers using the site factors and base station variables.

The equations used in MTCLIM to estimate the short wave radiation and vapour pressure deficit can be found in detail in the study carried out by Thornton and Running (1999).

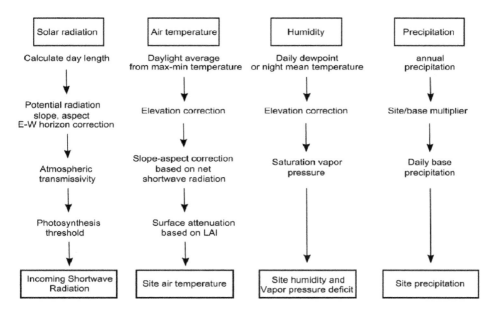

Figure 4-2 Flowchart of the subroutines of the MTCLIM model to estimate daily microclimate conditions in mountainous areas.

The specific ecophysiological parameters for each growth form of vegetation were derived from literature review and Chapter 3 (Table 4-3). This information included two important assumptions: (a) that there is a difference among the site and soil characteristics along an altitudinal gradient, and (b) that each of the growth forms of vegetation have different physiological characteristics. We considered it important to test these assumptions for further applications of the model in similar regions. Other additional information was derived from literature search for C3 grasses and calculated mean and standard deviation of each parameter (Table 4-3).

Table 4-3 Ecophysiology parameters for the main growth forms of vegetation: Tussocks (TU), Acaulescent Rosettes (AR) and Cushions (CU).

Parameter	TU			AR			CU			Estimations based on
	low	mid	high	low	mid	high	low	mid	high	
Turnover and mortality parameters										
Annual leaf & fine root turnover	2.017	1.825	1.667	1.807	1.383	1.120	1.299	1.116	0.979	Diemer (1998)
Annual whole plant mortality	0.34	0.22	0.19	0.34	0.22	0.19	0.34	0.22	0.19	Gill & Jackson (2000)
Current growth proportion	0.025	0.018	0.012	0.06	0.055	0.055		0.000128		TU: Scott(1961) AR: Diemer (1998) CU: Ralp (1978)
Allocation parameters										
Fine root C: leaf C	0.592	0.6	0.43	0.703	0.657	0.673	0.905	0.918	0.803	Section 3.3
Carbon to nitrogen parameters										
C:N of leaves	89.231	68.167	71.18	25.366	28.785	23.955	21.869	25.547	25.66	Section 3.3
C:N of leaf litter	141.9	108.4	113.2	40.34	45.78	38.1	34.78	40.63	40.81	Section 3.3
C:N roots	23.672	26.471	24.88	27.682	24.712	33.161	27.354	24.754	39.641	Section 3.3
Labile, cellulose, and lignin parameters										
Litter labile	0.506	0.523	0.512	0.501	0.546	0.473	0.413	0.44	0.403	Section 3.3
Litter cellulose	0.272	0.3	0.259	0.226	0.17	0.123	0.253	0.248	0.168	Section 3.3
Litter lignin	0.222	0.177	0.229	0.273	0.284	0.404	0.334	0.312	0.429	Section 3.3
Root labile	0.591	0.496	0.477	0.54	0.516	0.564	0.57	0.485	0.565	Section 3.3
Root cellulose	0.195	0.261	0.156	0.195	0.217	0.112	0.187	0.249	0.132	Section 3.3
Root lignin	0.214	0.243	0.368	0.265	0.267	0.324	0.243	0.266	0.303	Section 3.3
Canopy parameters										
Water interception (LAI^{-1} day^{-1})					0.0225					White et al. (2000)
Light extinction					0.48					White et al. (2000)
SLA (projected area basis) (m^{2}kg^{-1}C)	10.05	13.699	21.505	10.989	11.765	12.658	12.27	13.699	15.504	White et al. (2000)
Shaded/sunlit SLA					2					White et al. (2000)
All-sided: projected leaf area					2					White et al. (2000)
Leaf N in Rubisco (%)	1.764	1.837	3.01	0.307	0.373	0.334	0.295	0.385	0.438	Wullschleger (1993)
Maximum g$_s$ (mm s^{-1})		0.011			0.02			0.0217		TU: Kelliher et al. (1995) AR & CU: Cavieres et al.

				Reference
Cuticular conductance (mm s⁻¹)	0.00011	0.0002	0.00022	(2005) TU: Kelliher et al. (1995) AR & CU: Cavieres et al. (2005)
Boundary layer conductance (mm s⁻¹)	0.022	0.04	0.03	Nobel (1991)
Ψ_L start of g_s reduction (MPa)		-0.73		White et al. (2000)
Ψ_L complete of g_s reduction (MPa)		-2.7		White et al. (2000)
VPD start of g_s reduction (Pa)		1000		White et al. (2000)
VPD complete of g_s reduction (Pa)		5000		White et al. (2000)

C = carbon, N = nitrogen, LAI = leaf area index, SLA = specific leaf area, g_s = stomatal conductance, Ψ_L = leaf water potential, and VPD = vapour pressure deficit.

4.2.3 Model description

BIOME-BGC is an ecosystem process-based model which estimates fluxes and storage of energy (carbon, nitrogen and water) and mass for the vegetation and soil components (Thornton, 1998; Thornton et al., 2002). Figure 4-3 shows the schematic overview of the BIOME-BGC model, it uses site conditions, meteorological and eco-physiological data and Figure 4-4 shows the representation of the water, carbon and nitrogen dynamics in the model. The model has 34 parameters that can simulate daily fluxes of carbon, nitrogen and water for each biome. For C3 and C4 grasses it uses 21 of these parameters (Table 4-3) since there is no wood component in these vegetation types. It uses the Farquhar photosynthesis model (Farquhar et al., 1980) to estimate gross primary production (GPP), net primary production (NPP) and the net ecosystem exchange (NEE). The model performs a two-phase run. First a spin-up run that defines the equilibrium condition and the initial state variables of the system with the local eco-climatic characteristics. Later a transient or normal run which represents changes in the ecosystem's dynamics as it goes to a variation set of atmospheric CO_2.

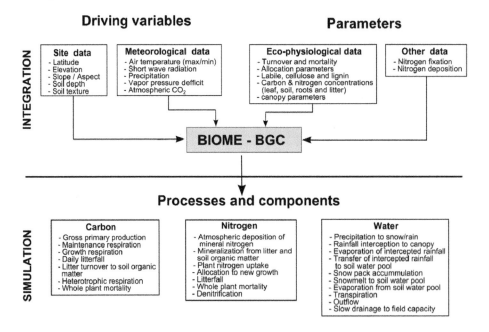

Figure 4-3 Schematic overview of the BIOME-BGC model.

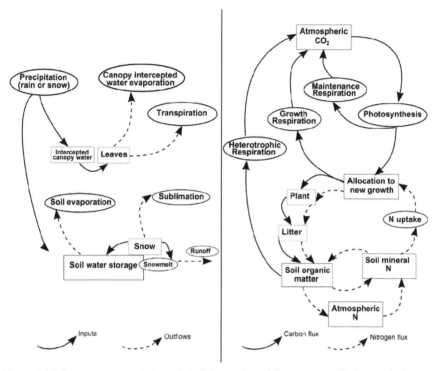

Figure 4-4 Schematic representation of a) Water pools and fluxes and b) Carbon and nitrogen dynamics in the BIOME-BGC model (Diagram taken from Lange (2010)).

The weather is the most important control on vegetation processes and it is the main driver for the ecosystem activity (Trusilova et al., 2009). The near-surface daily meteorological parameters required by BIOME-BGC are: maximum, minimum and daily average temperature (°C), total precipitation (cm), daylight average partial pressure of water vapor (Pa), daylight average shortwave radiant flux density (W/m^2) and daylength (s). A conceptual component of the major processes that BIOME-BGC uses to model the main growth forms of vegetation are detailed in Figure 4-5. We focused on both gross primary production (GPP) and stand hydrological processes, which were scaled to a per square meter basis and then extrapolated to the basin. These processes provide a means of estimating the spatial and temporal changes in carbon storage and water budgets for a better understanding of plant growth and production due to the availability of nutrient content and water in soil.

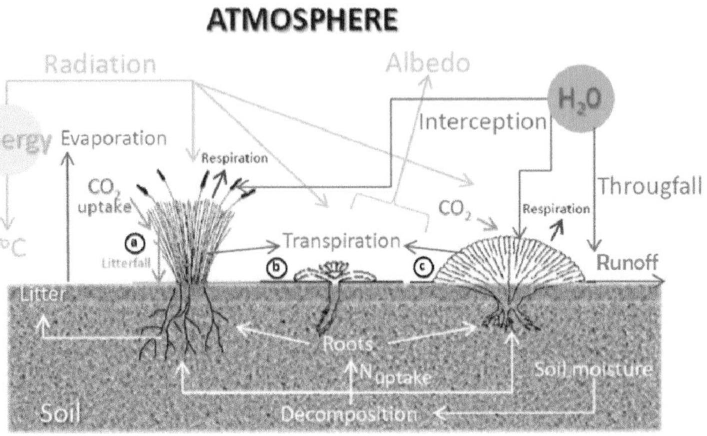

Figure 4-5 BIOME-BGC detailed model flow chart for a) Tussocks, b) Acaulescent Rosettes, and c) cushions.

4.2.4 Model simulation

The BIOME-BGC started with a spin-up simulation of a hypothetical steady state for the primary production in preindustrial conditions (Thornton and Rosenbloom, 2005). The spin-up simulation gets the model into equilibrium which typically reaches from 1200 to 3500 years. It uses the aforementioned daily meteorological records repeatedly along the run simulation, but does not take into account anthropogenic disturbances such as harvesting, replanting and grazing because they can overestimate the state variables. It was also performed with preindustrial parameterization using 280 ppm as a constant atmospheric CO_2 concentration (Rozanski et al., 1995) and 0.000389 kg N/m²/yr of nitrogen deposition (Phoenix et al., 2006) (Figure 4-6). Later, the model ran in normal mode with the same meteorological data, but this time with an annual variation of CO_2 (McGee, 2008) and 0.000667 kg N/m²/yr of nitrogen deposition (Phoenix et al., 2006). The nitrogen fixation for short grasslands was estimated at 0.0003 kg N/m²/yr for TU (Line and Loutit, 1973) and 0.00027 kg N/m²/yr for AR and CU (Cleveland et al., 1999). All biogeochemical processes were simulated within the vertical extent of a vegetation and its rooting system (Churkina et al., 2003).

The main assumptions used throughout the model simulation were that: a) the stationary behavior of ecophysiological parameters is fixed at each altitudinal; b) the system can be represented in a spatial grid cell distribution; c) there was non-interaction between cells (e.g. does not examine competitive dynamics across space). However, uncertainties that might emerge from the stationary-state assumption in the model logic (Carvalhais et al., 2008) can be assumed to be negligible due to low human intervention.

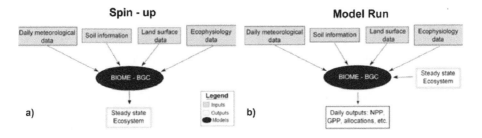

Figure 4-6 BIOME-BGC analysis scheme for a) Spin-up and, b) Model run simulation.

4.2.4.1 Gross primary production

The GPP provides the primary carbon input for the plant-level carbon cycle and represents the total carbon that can be distributed to different vegetation tissues based on the allometric relationships among them (Wang et al., 2009). The data required for GPP (kg C/m²) calculations include daily incoming radiation (W/m²), vapor pressure deficit (Pa), and air temperature (°C). Further inner processes can be found in detail in the research done by Thornton (Thornton, 1998). GPP was extrapolated for the entire basin based on the pixel classification of the growth form coverage in different altitudinal ranges (Refer to Figure 4-1).

One-way analysis of variance (ANOVA) was applied to test the differences in GPP among growth forms of vegetation and altitudinal ranges. The ratio used for the contribution of each growth form to the total GPP at each altitudinal range was calculated as follows:

$$Ratio = \frac{GPP_{GF}}{GPP_R} \tag{4-1}$$

where GPP_{GF} is the GPP estimated in each growth form of *páramo* vegetation (AR, TU, CU) and GPP_R is the GPP estimated in each altitudinal range (Low, Mid, High).

We decomposed the frequency domain representation of the GPP simulation at each altitudinal range by performing a Fourier series analysis. The difference between the Fourier series and the GPP simulated was called signal. The signal correction was applied throughout the analysis and was later compared to the average GPP. Average GPP was calculated as the product of the above-ground biomass (kg/m²) and the carbon concentration (%) in the plant material, from now on called ground-based GPP (Eq.4-2). The values for $Biomass_{above-ground}$ were taken from Table 3-4.

$$GPP_{Ground-base} = Biomass_{above-ground} \times Carbon\ (\%) \tag{4-2}$$

4.2.4.2 Soil and stand hydrological processes

Climate and elevation play an important role in determining the soil properties as shown in section 3.3 and therefore soil water content. The open and porous structure of the soil at higher elevations increases the infiltration capacity, thus limiting nutrient provision and root growth. The initial water content as a proportion of saturation is 0.82 for low and mid elevation and 0.55 for high elevation (Buytaert et al., 2005a). The model assumes constant soil temperature throughout the routing zone and simulates a soil temperature profile in the upper 5cm, which is the zone that is coupled with the atmosphere (Trusilova et al., 2009).

In BIOME-BGC, the Penman-Monteith equation (Monteith, 1973) is used to calculate the soil water evaporation, evaporation from the canopy interception and the transpiration from the leaves. Additionally, the model computes the storage as water that is kept in the soil and the outflow as the water that leaves the system. Estimated soil evaporation takes into account the decrease in water availability between rain events by reducing the evaporation rate accordingly (Taiz and Zeiger, 2006). Canopy evapotranspiration sums leaf transpiration and evaporation from plant surfaces. Canopy water interception for all *páramo* vegetation is 0.0225 (1/LAI/d) (White et al., 2000a) which accounts for this vegetation's ability to absorb water through their stomata.

The model updates all the state variables and checks the balance function to control for the principle of the Conservation of Mass (Thornton, 2003; Trusilova and Churkina, 2008; Trusilova et al., 2009; Uhlenbrook, 1999; Wang et al., 2009). The water balance in time (t) is calculated as follows (Eq.(4-3)),

$$Balance\ (t) = input\ (t) - outflow(t) - \frac{dS\ (t)}{dt} \tag{4-3}$$

where *input* is the precipitation; *outflow* is the sum of runoff, soil evaporation, canopy evaporation and leaf transpiration; and $\frac{dS\ (t)}{dt}$ is the change in soil water storage and the water in the canopy.

4.3 Results

4.3.1 Model parameterization and calibration

The model parameterization was presented and it is found in (Table 4-3). The table showed all the parameters for the three vegetation growth forms considered within this study. The parameter values presented here are the mean and the standard deviation, showing the variation along an altitudinal gradient.

Data for grasses was scarce; thus the available information was reported without discrimination between C3 and C4 grasses. Our parameterization and calibration processes rely on the statistical analysis of key parameters derived from in-situ measurements in order to produce a significant reduction in the uncertainty of GPP simulations.

4.3.2 Estimation of GPP

From the ANOVA analysis (described section 4.2.2), the GPP estimations were higher at low and mid altitudinal ranges (Low-High $p < 0.001$, Mid-High $p < 0.01$). This was also tested among growth forms of vegetation (AR-TU $p < 0.001$, CU-TU $p < 0.001$). Figure 4-7 shows the boxplots of the GPP among altitudinal ranges and the three forms of vegetation (used in the ANOVA analysis).

Figure 4-7 Boxplots (95% confidence interval) of GPP variation comparing a) Altitudinal ranges (Low, Mid, High), b) Growth forms of vegetation (TU=tussocks, AR= acaulescent rosettes, CU= cushions).

4.3.2.1 Contribution of each growth form to the total GPP

The GPP ratio (Eq. 4-1) is shown in Figure 4-8 for each altitudinal range and the three growth forms of vegetation. At low altitudinal range, the contribution of tussocks to the total GPP is highly significant (around 70%), three times the contribution of acaulescent rosettes (Figure 4-8a). Cushions do not have a significant contribution at any altitude (Figure 4-8).

Variations in the GPP ratio showed fluctuations with almost similar values in 2010 (Figure 4-8c), which was a dry hydrological year. The similarity reflected the ambiguous responses and uncertainty in the model forcing fields when vegetation is located in the upper line close to the permanent snow.

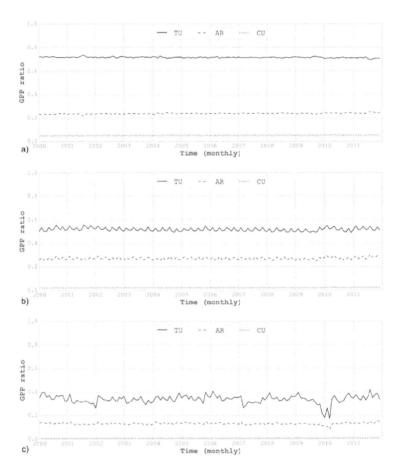

Figure 4-8 Ratio between the growth forms (TU=tussocks, AR= acaulescent rosettes, CU= cushions) at each altitudinal range a) Low, b) Mid, and c) High.

4.3.2.2 GPP accumulation along the altitudinal gradient

We considered two monthly events that include the maximum and minimum GPP values and correspond to the high and low production seasons, respectively. The monthly contribution of GPP for the basin during a high production season is around 2295 kg C, which is more than 3.5 times the GPP estimated during a low production season (620 kg C) (Figure 4-9). Figure 4-9 shows that the greatest contribution of GPP occurs in the first altitudinal range (4100-4200 m a.s.l.). Above 4500 m a.s.l., the contribution of GPP becomes smaller.

The production of GPP in the basin is related to the percentage of coverage of each growth form and their capacity of energy production. Tussocks not only have larger coverage (around 60%) (Refer to Figure 4-1) at low and mid elevations but they are also capable of producing more GPP, as seen in the previous results (Figure 4-8).

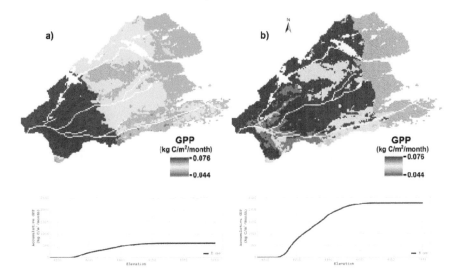

Figure 4-9 Scenarios of a) low and b) high GPP (kg C/m²/month) along the altitudinal gradient.

4.3.2.3 BIOME-BGC GPP vs ground-based GPP

Figure 4-10 provides more detailed insight into the different GPPs per altitudinal range and their own variation in the min-max threshold range. The signal calculated from the BIOME-BGC GPP fluctuates around the average value of the ground-based GPP. If there was a constant above-ground biomass and carbon concentration in the plant material, the maximum and minimum values of the ground-based GPP were representative of the upper and lower thresholds expected of GPP. The ground-based GPP estimation did not involve changes of precipitation, temperature, nitrogen limitation, radiation, and therefore there was no variation of the monthly GPP in each altitudinal range (Figure 4-10). BIOME-BGC's GPP estimations for the mid altitudinal range showed less fluctuations outside the threshold limits. However, for low and high elevations, the GPP was systematically over- or under-estimated by the model, being spread above and below the thresholds of ground-based GPP.

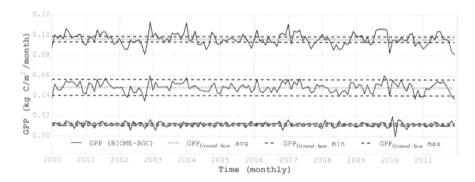

Figure 4-10 Maximum, minimum and average values of $GPP_{Ground-base}$ and GPP (simulated with BIOME-BGC for each altitudinal range: a) Low, b) Mid, and c) High.

4.3.3 Water Balance

Figure 4-11 shows the results for tussocks at mid altitudinal range at a monthly basis for the 12-year period. Simulation runs are presented for the water budgets, precipitation, evaporative processes, which is the sum of soil and canopy evaporation and leaf transpiration and the change in storage that is represented as surplus (SM Surplus) or deficit (SM Deficit) of soil moisture. In most monthly water budgets, there was more precipitation than evaporative processes represented by the soil moisture surplus for all growth forms of vegetation along the altitudinal gradient, which means that there is more water flowing into the system than water leaving accounted as soil evaporation, canopy evaporation and leaf transpiration. However, there are a few months where the evaporative processes slightly exceed the precipitation producing a soil moisture deficit. This soil moisture deficit mostly in the second semester of the year is directly linked to the high average temperatures registered from October to January and lower average precipitation values from August to November (Figure 2-5). It must be noted that the soil moisture surplus includes the soil moisture recharge that happens right after a deficit. In the same way the soil moisture deficit includes the soil moisture utilisation that takes place after a surplus month. Exceptionally in 2010 due to the low monthly precipitation registered in the basin, there is a soil moisture deficit from at least seven months. The results of soil moisture surplus showed higher values in the first quarter of each year due to high values of precipitation. The water balance is nearly zero throughout the simulation (Figure 4-11 caption b).

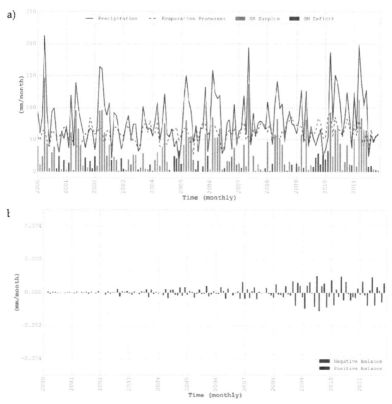

Figure 4-11 a) Representation of water budgets, precipitation, evaporative processes (sum of soil evaporation, canopy evaporation and leaf transpiration), and soil moisture (deficit and surplus), and b) Water balance difference for tussocks at mid altitudinal range.

4.4 Discussion

4.4.1 Current and long-term carbon dynamics

GPP simulation of the growth forms represents a large group of *páramo* grasslands, while inheriting the trends, sensitivities, strengths and limitations of the BIOME-BGC. *Páramo*, as any other ecosystem, deals with a range of natural and anthropogenic disturbances moving it far from an equilibrium stage. Unfortunately, simulations ignore these impacts leading them to an overestimation of the state variables. The study area lies within an ecological reserve, which is considered as a low disturbed ecosystem. Therefore the differences between the modeled and the ground-based GPP are mainly associated with the changes in climatic conditions. The results suggest that the magnitude of ecosystem's response to environmental changes is closely related to the site's climate variability and site-specific features

(Churkina et al., 2003). The decrease in GPP towards higher altitudes suggests a strong combined effect from the reduction of vegetation cover and the variation of its parameters along an altitudinal gradient. This is consistent with the knowledge that, high altitudes are characterized by less favorable climatic conditions for vegetation and that non-vegetated soils have high nutrient losses (O'Connor et al., 1999) to which few plants can survive or adapt (Sklenar and Ramsay, 2001). Thus, not considering the role altitudinal gradient in modeling GPP might result in underestimation is likely to happen.

The *páramo* ecosystem is further characterized by low productivity and low growth rate of the vegetation. These characteristics increases and lengthen carbon accumulation in the geobiochemical carbon cycle. Therefore, the *páramo* vegetation and the soil can be important sinks for atmospheric CO_2, if appropriately managed. But, carbon stored in both soil and vegetation must be decycled and removed from the geobiochemical carbon cycle to be effective against climate change. Several poor management practices, such as the progress of the agricultural frontier and the change in the land use can endanger endemic vegetation and consequently, carbon stored in plant material.

4.4.2 Water balance

Model simulations of the stand hydrological processes tested water interactions and processes as well as the acceptability of the model's simplified representation of the water cycle. Unfortunately, the BIOME-BGC has limitations in the conceptualization of the soil component; it uses a simplified approach of a 1D bucket model that limit the possibility for the model to capture the soil profile dynamics and soil water capacity.

The evaporative processes of canopy evaporation of intercepted water and transpiration during photosynthesis are well represented in the model since it takes into account the leaf conductance to water vapour, which relates the stomatal conductance to water and the leaf conductance to sensible heat and it varies across species and their response to water stress. However, there might be room for improvement since all parameters related to the reduction of stomatal conductance were obtained from a general type of grasses in literature and not specifically for the main growth forms of vegetation discussed in this study. A better estimation of these parameters will improve the assessment of the evaporative processes particularly for tussock grasses, which are characterized by having low evaporation regardless of the high evaporative force of the intense ultra-violet radiation of the equatorial high mountains (Buytaert et al., 2004).

The soil moisture surplus and deficit calculated cannot provide information about the subsurface hydrological processes and therefore an accurate calculation of soil moisture changes in the soil compartment due to diffusion, percolation, and root water uptake. We hypothesize an underestimation of the system's water storage capacity, specifically water retention in the *páramo* soils. Water stored in the soil is a function of saturated soil water and field capacity soil, both based on the soil texture and depth (White et al., 2000a). These were different from low to high elevation ranges but not significantly different from plant to plant. The *páramo* is a peculiar ecosystem, considered as one of the fastest evolving biodiversity hotspot worldwide (Madriñán et al., 2013). Because in pristine conditions it acts as a sponge and water retention in its soils reach 200% its own dry weight (Cañadas Cruz, 1983). Eventually, excess water is gradually restored to the ecosystem at lower elevations (Gómez Molina and Little, 1981), but this is not accounted for in the model. Undoubtedly, the underestimation in the water storage capacity justifies adapting the model's parameter estimation to use multiple soil layers for calculating the soil temperature. Currently, only one soil pool is used, which removes the possibility for the model to capture the soil profile dynamics and limits the saturated soil water capacity, possibly resulting in overestimated runoff.

4.4.3 Challenges in the use of BIOME-BGC

The study runs a dimensional model that represents a point with all fluxes and stocks scaled on a per square meter basis. Then extends it over a spatial context that covers the classification of vegetation growth forms in the basin. Each cell represents a distinct model run and does not interact with other cells, which also means that the model does not examine the competitive dynamics across the space and shading of other plants differing in height growth. Not all the plants grow, breathe, die and decay at the same time. The model uses leaf longevity to assess plant mortality. However, the model requires adaptation in order to take into account the plant's onset and senescence that represents the reality of the *páramo* ecosystem. This lies in the context of the assumptions stated earlier, where the model does not examine competitive dynamics across space.

The BIOME-BGC model representation cannot provide a detailed impact analysis for the vegetation due to hydro-meteorological variation. This might be explained by the slow growth of the equations representing the *páramo* ecosystem. The temporal resolution chosen was daily even though there are some sub-daily ecosystem dynamics (sun spots, shade and clouds, gust of wind) (Trusilova et al., 2009). This is because there are no model validation techniques accurate enough to provide information about the major carbon components (C uptake, plant respiration, soil microbial respiration) and hydrological cycle components (vertical and horizontal precipitation, transpiration, evaporation, evapotranspiration, infiltration and redistribution) (Churkina et al., 2003) at a sub-daily time scale. Consequently, it was difficult to validate every aspect of the ecosystem model at a higher temporal resolution because not all components can be measured in the field. However, the impact of this limitation is somewhat reduced because the Andean ecosystem has no strong temperature seasonality and it is not marked with summer and winter season. Thus, the *páramo* vegetation is continuous throughout the year.

In the Andean mountain region, the evaporation and transpiration are the most important hydrologic components and indicators of water availability and ecosystem productivity. Therefore, an energy balance model should be implemented to properly estimate daily and monthly evapotranspiration processes. It is also important to include input of water that is not in the form of precipitation. The interception of fog and mist are very common in the region due to the orographic uplift caused by the Andes. Although fog and mist only represent an unquantified and small percentage that mostly comes into contact with tall tussock grasses and arbustive vegetation, it should be neglected. Their interception potentially influences the primarily soil moisture conditions in the upper soil layer and thus ecosystem function (Nagy et al., 2010).

4.5 Conclusions

The model simulations with BIOME-BGC show a distinctive contribution of vegetation growth forms to total GPP along an altitudinal gradient as well as a gradually decreasing pattern in the GPP estimation towards higher elevations. These dynamics are seemed to depend not only on the eco-physiological parameters of the growth forms but also on the variation of climatic drivers in the catchment. Tussocks have proven to be the leading growth form of vegetation fixing the largest part of the carbon during photosynthesis in the *páramo* ecosystem.

A good understanding of the ecological process of the *páramo* vegetation is essential for successful local to regional and even global ecosystem scale applications of the model since plant-specific information is usually missing at those scales.

Some structural developments of BIOME-BGC with respect to the water fluxes and plant mortality are important next steps to better model *páramo* ecosystems. Further understanding of carbon sequestration details and stand hydrological processes are necessary for a successful application of BIOME-BGC to this type of ecosystem.

This study identified the key vulnerabilities in soil-vegetation interaction at higher altitudes and their potential feedback to the biosphere. As such, this research is well aligned with the priorities in Chapter 4 of the IPCC Report (Fischlin et al., 2007), which are 'to improve representation of the interactive coupling between ecosystems and the climate system'. It does so by reducing the uncertainties in GPP estimations and threshold responses by using more realistic characteristics of the ecosystems.

BIOME-BGC does not allow non-precipitation water inputs. This is rather, unfortunate not only for the *páramo*, but also for other types of ecosystems such as tropical rainforest and lower montane cloud forests, where this "horizontal rain" contributes some 5 to 20% of ordinary (vertical) rain (Ataroff and Rada, 2000; Bruijnzeel and Proctor, 1995). A reduction in water input uncertainty requires developing an additional variable representing this 'horizontal rain' in the modelling system.

There are limitations in the estimation of the soil moisture since the soil block is represented as a 1D bucket model that limit the possibility for the model to capture the soil profile dynamics and soil water capacity. Further research should focus on the development of a multilayer soil water module to estimate a more accurate soil property of moisture content. A correct description of soil moisture will improve the carbon and water cycle of the *páramo* vegetation.

Acknowledgements

Thanks also to Aline Saraiva Okello, Juan Carlos Chacón, Patricia Trambauer for the tips and help in the coding troubleshooting, and Kirstin Conti who proofread the manuscript from which this chapter is based on.

Imagination is more important than knowledge (A. Einstein).

5

ANALYSIS OF THE RELATIONSHIP BETWEEN CLIMATE VARIABLES AND GROSS PRIMARY PRODUCTION USING DATA DRIVEN TECHNIQUES

As one of the main areas of carbon cycle and climate change studies, water and CO_2 relations are of great significance in the estimation of gross primary production (GPP). Various process-based models have been set up to estimate the GPP based on mathematical representation of biological, physiological and ecological processes aiming to represent real conditions. However, they often end up increasing the complexity and computational processing power due to the large number of physical equations that need to be solved. Computation time becomes an important matter in the simulation of multiple scenarios using models for long periods of time (e.g. climate projections). This chapter evaluates the performance of four DDMs as surrogate models, i.e. linear regression method (LRM), model trees (MT), instance-based learning (IBL) and artificial neural networks (ANN). The aim is to explore whether DDMs are useful to evaluate the complex relations between processes, and which are the main climatic drivers for gross primary production.

This chapter is based on:

Minaya V., Corzo G., Solomatine, D., and Mynett, A. E.: Data-driven techniques for modelling the gross primary production of the *páramo* vegetation using climate time-series data, application in the Ecuadorian Andean region, Ecological Informatics, In revision, 2016.

Minaya, V., Corzo, G., Van der Kwast, J., Galarraga-Sanchez, R., and Mynett, A. E.: Classification and multivariate analysis of differences in gross primary production at different elevations using BIOME-BGC in the *páramos*; Ecuadorian Andean Region. In: Revista de Matemática: Teoría y aplicaciones Vol.22, No.2, 2, CIMPA, San Jose - Costa Rica. ISSN: 1409-2433, 2015.

5.1 Introduction

There is a growing interest on the estimation of terrestrial gross primary production (GPP) around the world, whether these ecosystems are acting as sources or sinks of carbon and how they potentially can contribute to the effects of climate change (Prentice et al., 2000). The GPP is the amount of carbon dioxide that is taken up by the plants during photosynthesis and it is allocated for plant biomass production and respiration (Gough, 2011). GPP supports human well-being since it is the basis for food, fiber, wood production, and fuel. Additionally, GPP is one of the largest global CO_2 flux that controls several ecosystem functions (Beer et al., 2010); e.g. land-atmosphere interactions and carbon sequestration. Many process-oriented models have been proposed to deal with the complex interactions; however, in some cases, these models include a number of scientific hypotheses done for a particular ecosystem that might end up in an erroneous generalization of another ecosystem. An attempt to estimate this different at large grid cells was done by Minaya et al. (2015b; 2016), showing an error of 23% in average if no spatial heterogeneity is considered in the *páramos*. The *páramo* ecosystem is a complex 'hot spot' mountain ecosystem that holds a great amount of biodiversity and unique ecological processes, thus providing important ecosystem services in terms of hydrological regulation and carbon storage.

There is a growing interest in the estimation of terrestrial gross primary production (GPP) of ecosystems due to their role as sources or sinks of carbon and their contribution to the effects of climate change (Prentice et al., 2000). The GPP is the total amount of energy produced by the plants during photosynthesis and used for biomass production and respiration (Gough, 2011). GPP supports human well-being since it is the basis for food, fibre, wood production, and fuel. Additionally, GPP is one of the largest global CO_2 fluxes that controls several ecosystem functions (Beer et al., 2010); e.g. land-atmosphere interactions and carbon sequestration. Many process-oriented models have been proposed to deal with these complex interactions; however, in some cases, these models include a number of scientific hypotheses adopted for a particular ecosystem that might end up in an erroneous generalization when used for another ecosystem. An attempt to estimate this difference at large grid cells was undertaken by Minaya et al. (2015b; 2016), showing an error of 23% on average if no spatial heterogeneity is considered. The mentioned studies have been carried out for *páramos*, a complex 'hot spot' mountain ecosystem that holds a great amount of biodiversity and unique ecological processes, thus providing important ecosystem services in terms of hydrological regulation and carbon storage.

Several vegetation and ecophysiological models have attempted to recreate the variation of GPP and evaluate the system behaviour (Cramer et al., 2001; McGuire et al., 2001). Validations have been carried out using carbon exchange monitoring measurements and also above and belowground biomass estimations, if available (Belgrano et al., 2001; Hilbert and Ostendorf, 2001; Jung et al., 2007). However, there are some modelling issues that are difficult to resolve and which in most of the cases have been neglected or generalized leading to high uncertainties (Moorcroft, 2006; Morales et al., 2005). These refer to carbon monitoring in a consistent manner, approximations of nonexistent data, homogenization of plant functional types, static model parameters and site descriptors unchanged within an altitudinal gradient. On top of this, distributed process-oriented models can be computationally expensive: they typically have more than 30 parameters only to describe the vegetation processes.

In an attempt to reduce computational load during the model use, the use of surrogate models, i.e. simplified models (typically, data-driven) of process models, could be an alternative (Koziel and Leifsson, 2013; Regis and Shoemaker, 2013). However building them also requires building the process models first (with all the limitations mentioned above), and generation of large data sets for training the surrogate model requires multiple model runs leading to a serious computational effort as well; this is however done once, off-line, and experiments with the resulting model do not require much time.

Data driven models (DDM) are constructed to represent complex interactions and allow data analysis, identification of trends and feasible predictions (Belgrano et al., 2001; Hilbert and Ostendorf, 2001; Papale and Valentini, 2003; Zhang et al., 2007). Basically a DDM is a (non-linear) statistical model describing the relationships between the input and output variables characterising the studied system. DDMs depend much less on theoretical assumptions, and in this regard are complementary to the process based models. DDMs have limitations: they depend on the quality of the used data set and cannot generalise well for different conditions and use cases.

In ecological modelling numerous applications of a wide variety of DDM techniques have been reported, showing that it is possible to represent complex relationships which are not clearly explained by physically- or biologically-based considerations. In spatial dynamics, for instance, cellular automata and artificial neural networks have been applied for primary production (Anav et al., 2015; Belgrano et al., 2001; Scardi, 1996), carbon dioxide uptake and other carbon fluxes (Beer et al., 2010; Jung M et al., 2011; Papale and Valentini, 2003; Xiao et al., 2014), radar forecasting (Li et al., 2013). DDMs have been also used for algae growth (Chen and Mynett, 2006; Li et al., 2010; Recknagel et al., 1997; Scardi, 1996), classification of landscape types (Brown et al., 1998; Zhang et al., 2007), distribution of vegetation (Hilbert and Ostendorf, 2001; Linderman et al., 2004) and hydrologic modelling for climate change scenarios (Corzo et al., 2009; Elshorbagy et al., 2010).

Looking specifically at terrestrial primary production, several studies have compared the use of data-driven methods such as multiple regression models and artificial neural networks (ANN) for a particular time frame (Jung et al., 2008; Papale and Valentini, 2003; Paruelo and Tomasel, 1997; Vetter et al., 2008). However, limited references are available in the knowledge of the author and none of them have considered various time frames and a broader comparison of several DDMs. By comparing various time frames is possible to enhance the understanding of how influential is the changes of temporal resolution and the selection of the precise meteorological variables for the selection of the most adequate DDM that generate the minimum error.

This study evaluates the performance of data-driven model (DDM) techniques to discover the complex interactions present between GPP and meteorological variables at various time frames. The GPP results from the BIOME-BGC model from Chapter 4 were used as reference biomodel. Four DDMs where built as surrogate to simulate the GPP obtained from the biomodel, namely: linear regression method (LRM), model tree (MT), instance-based learning (IBL) and artificial neural network (ANN) model.

5.2 Methods and data

The case study is the same described earlier in Chapter 2 (sections 2.2 to 2.4). As demonstrated earlier in Chapter 3, the growth forms had large differences in their carbon, nitrogen concentration and main ecophysiological characteristics along altitudinal gradients (Minaya et al., 2015a). In this regard, the analysis will keep the same analysis at three elevations (R1: 4000-4200 masl; R2: 4200-4400 masl; R3: 4400-4700 masl).

5.2.1 Data

Meteorological data such as precipitation (PRE), short wave radiation (SWR), vapor pressure deficit (VPD) and temperature average (TAVG), minimum (TMIN) and maximum temperature (TMAX) were the same ones used for the simulation of gross primary production (GPP). This was discussed in Chapter 4, where GPP was estimated using BIOME-BGC (BioGeochemical Cycles), which is an ecosystem process model that estimates fluxes and storage of energy, water, carbon and nitrogen for soil and vegetation of terrestrial ecosystems (version 4.2 Thornton, 1998; Thornton et al., 2002).

GPP is defined as the total amount of CO_2 that is fixed by the plants through photosynthesis and it has proved to be a good indicator of ecosystem's health, high GPP means high amount of CO_2 sequestration on the region and low values mean plant decay and organic matter decomposition. The model parameterization was done in Chapter 4 that relied on statistical analysis of key parameters derived from in situ measurements in order to reduce significantly the uncertainty of GPP simulation. The GPP for each altitudinal range took into account the percentage of coverage of each growth form.

5.2.2 Data preparation

All climatic variables were aggregated at a monthly scale, and tested for simple monotonic trend using Mann-Kendall test (Gilbert, 1987; Kendall, 1975; Mann, 1945) over 12-years time, there was no trend discovered (p<0.05). Input variable selection (IVS) was used to decide on the most relevant and potential environmental model inputs (see e.g. Galelli et al. (2014)).

Data from three altitudinal ranges was used and grouped at various time frames (daily, weekly, bimonthly and monthly). The values of precipitation and GPP were aggregated while for other climatic variables an average function was used.

5.2.3 Methodology

The methodology includes two main steps. At the first step, model-free analysis using criteria of correlation coefficient (CC) and average mutual information (AMI) between the climate time-series data and the GPP is carried out. At the second step, model-based analysis is used to test four different combinations of input variable sets, in which the lowest root mean square error (RMSE) leads to choose the best DDM. These four combinations are obtained following backward elimination defined by Blum and Langley (1997), where the search starts at a full selection of climatic variables and then variables are removed based on the performance and expert judgement. Bimonthly and monthly data sets were used to train LRM, MT, IBL and ANN model. Normalized root mean squared error (NRMSE) was used as an indicator of the surrogate model performance. When looking at the bimonthly and monthly time frames, the lowest NRMSE values were shown in the monthly time frame. All steps carried out are summarized in Figure 5-1.

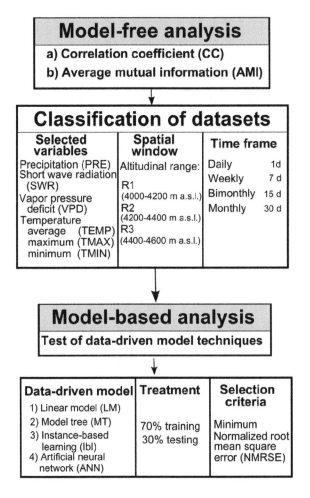

Figure 5-1 Sketch representation of the steps followed for the selection of surrogate data-driven model

5.2.4 DDM set-up

For the development of surrogate DDMs, data was split in two data sets, 70% for training (from August 2003 to December 2011) and 30% for testing (from January 2000 to July 2003), following Elshorbagy et al. (2010). Table 5-1 shows that statistical properties of both data sets are similar.

Table 5-1 Statistical properties of the training and testing data sets for all input and output variables.

Variable	R1 Training		R1 Testing		R2 Training		R2 Testing		R3 Training		R3 Testing	
	μ	σ	μ	σ	μ	σ	μ	σ	μ	σ	μ	σ
GPP	0.006	0.002	0.005	0.002	0.004	0.001	0.004	0.001	0.003	0.001	0.002	0.001
PRE	2.6	4.1	2.6	4.3	3.7	5.8	3.8	6.5	3.1	4.6	3.2	5.0
SWR	415.7	94.8	420.3	102.1	420.5	95.1	424.7	102.3	361.7	87.1	385.7	90.5
VPD	396.4	125.7	370.1	124.6	318.4	117.1	296.3	118.0	292.6	76.8	308.1	76.1
TEMP	7.6	1.6	7.2	1.5	6.4	1.5	6.2	1.5	4.8	1.4	4.8	1.3
TMAX	10.2	2.2	9.6	2.1	8.6	2.0	8.2	2.1	7.0	1.8	7.1	1.7
TMIN	0.8	1.7	0.7	1.8	0.7	1.8	0.9	1.8	-1.0	0.9	-1.4	0.8

5.2.4.1 Linear regression method and Model Tree set up

The LRM and MT were built using the WEKA software (Witten and Frank, 2000). MT was built using the MP5 algorithm. It progressively splits the data set trying to ensure low standard deviation in subsets, and eventually generates several linear regression models for the resulting subsets (in the tree leaves). The minimum number of instances per leaf was set to 4.

5.2.4.1 Instance-based model set up

For instance-based learning (IBL) we used k-nearest neighbour algorithm for regression which was implemented in Matlab. The classification of the GPP was based on the Euclidean distance and calculated for each of the input vectors of the training set. The k-nearest neighbour algorithm weighted each of the k neighbours X_i based on their distance to the query point Xq, calculated as:

$$f(q) = \sum_{i=1}^{k} W_i f(X_i) / \sum_{i=1}^{k} W_i \tag{5-1}$$

where W_i is a function of the distance between X_q and X_i. The weighing scheme tested was linear:

$$wi = 1 - d(Xq, Xi) \tag{5-2}$$

The number of neighbours chosen was based on the minimum root mean squared error (RMSE) of the GPP from biomodel and the surrogate model. Since for large data sets IBL can be time-consuming, we limited our search to 30 nearest neighbours.

5.2.4.2 ANN model set up

We employed an artificial neural network (ANN) with a multilayer perceptron (MLP) structure with the logistic function in all nodes. The number of nodes in the ANN model structure was determined by exhaustive model optimization, varying the number of nodes from 5 to 20, and selecting the best performance (root mean square error, RMSE) on testing data. Training was carried out by the back propagation algorithm. To prevent so-called network paralysis output data were normalized to a range [0,1]. The non-linear transfer logistic function was bind between 0.1 and 0.9 to avoid negative GPP values during the extrapolation. The ANN optimization algorithm used a regularized error (msereg) to reduce the overfitting and in general to improve the generalization of the neural network, it was calculated as:

$$msereg = Y\,mse + (1 - Y)msw \tag{5-3}$$

$$mse = \frac{1}{N}\sum_{i=1}^{N}(e_i)^2 \tag{5-4}$$

$$msw = \frac{1}{n}\sum_{j=1}^{n}w_j{}^2 \tag{5-5}$$

where Y is the performance ratio, mse is the mean square error, msw is the mean square weight, e_i are the network errors and w_j are the network-weights..

5.2.5 Comparing model performance

We measured the performance of the techniques mentioned above: linear regression method (LRM), model tree (MT), instance-based learning (IBL) and artificial neural network (ANN). The model comparison was based on the normalized root mean square error (NRMSE) on the testing data set.

Taylor diagrams (Taylor, 2001) were used to compare the performance based on Pearson correlation, standard deviation and root mean square error (RMSE) of the surrogate models. These diagrams display pattern statistics; the radial distance from the centre of the graph is proportional to the standard deviation of a pattern. The

Pearson correlation between the two models is given by the azimuthal position, to the right positive correlations and left negative correlations.

Figure 5-2 shows the time-series data of GPP and meteorological values for monthly time frame at altitudinal range R2. These data were used in building all DDMs.

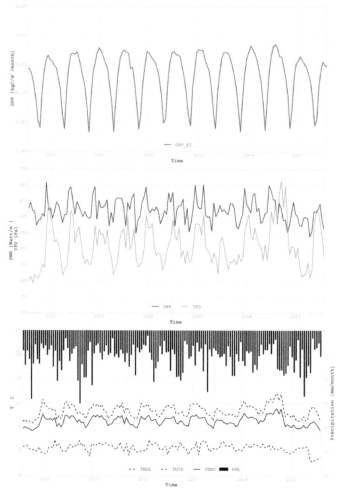

Figure 5-2 Monthly time-series data of GPP and meteorological variables for 12 years (2000 - 2011) at altitudinal range R2 (4200-4400 m a.s.l.)

5.3 Results

5.3.1 Model-free IVS

The results of the application of the model-free approach for all altitudinal ranges are presented in Table 5-2, the input variables are all the climatic drivers (PRE, SWR,

VPD, TEMP, TMAX, TMIN) that were considered important and linked to the variation of GPP. The highest values for the various time frames are shown in bold, bi-monthly and monthly time frames were used later for the analysis with different combinations of input variable sets as shown in Table 5-3.

Table 5-2 Model-free technique using criteria of correlation (CC) and average mutual information (AMI) between the climatic time-series data of precipitation (PRE), short wave radiation (SWR), vapor pressure deficit (VPD), mean temperature (TEMP), maximum temperature (TMAX), minimum temperature (TMIN) and gross primary production (GPP). Twelve years of time-series data (2000-2011) aggregated in four time frames: daily (1d), weekly (7d), bimonthly (15d) and monthly (30d) at three main altitudinal ranges: R1: 4000-4200 m a.s.l.; R2: 4200-4400 m a.s.l.; R3: 4400-4700 m a.s.l. Significant correlations ($p<0.05$) and high values of AMI are indicated in bold.

Altitudinal Ranges	Input variables	Output variable	Model-free							
			CC				AMI			
			1d	7d	15d	30d	1d	7d	15d	30d
R1	PRE	GPP	-0.16	-0.24	-0.30	-0.39	0.041	0.080	0.179	**0.336**
	SWR		0.34	0.40	**0.46**	**0.56**	0.144	0.146	0.232	**0.419**
	VPD		0.38	0.44	**0.46**	**0.49**	0.164	0.196	0.274	**0.479**
	TEMP		0.36	0.40	0.41	0.42	0.159	0.218	0.256	**0.413**
	TMAX		0.40	0.44	**0.45**	**0.47**	0.184	0.220	**0.308**	**0.436**
	TMIN		-0.11	-0.09	-0.10	-0.12	0.052	0.074	0.127	0.290
R2	PRE	GPP	-0.14	-0.21	-0.27	-0.34	0.017	0.075	0.154	**0.518**
	SWR		0.31	0.38	0.44	**0.53**	0.093	0.125	0.212	**0.561**
	VPD		0.38	0.44	**0.47**	**0.52**	0.127	0.210	0.258	**0.696**
	TEMP		0.40	**0.47**	**0.49**	**0.50**	0.158	0.238	**0.327**	**0.706**
	TMAX		0.42	**0.49**	**0.51**	**0.54**	0.168	0.250	**0.326**	**0.683**
	TMIN		-0.07	-0.04	-0.04	-0.05	0.029	0.067	0.126	**0.543**
R3	PRE	GPP	-0.11	-0.10	-0.10	-0.11	0.032	0.028	0.152	**0.483**
	SWR		0.31	0.32	0.34	0.38	0.110	0.073	0.227	**0.539**
	VPD		0.29	0.27	0.27	0.27	0.097	0.060	0.229	**0.585**
	TEMP		0.26	0.24	0.23	0.22	0.101	0.054	0.250	**0.637**
	TMAX		0.28	0.26	0.25	0.25	0.106	0.069	0.230	**0.600**
	TMIN		-0.04	-0.04	-0.04	-0.06	0.030	0.040	0.219	**0.556**

Correlations were checked also with various time frames at the three elevations (Table 5-2), the highest correlations appear when data was aggregated in bimonthly and monthly time frames. For the low and mid elevations (R1 and R2, respectively), the GPP is positively correlated to SWR (rho > 0.50, $p<0.05$) to VPD (rho > 0.60, $p<0.01$), mean and maximum temperature (rho >0.55, $p<0.05$ and rho >0.60, $p<0.01$ respectively). Conversely, precipitation and minimum temperature showed low values of correlation and AMI displaying a little direct relationship with the GPP. For higher elevations (R3) correlation was very low but it showed high AMI values, we assumed there was no good association between variables. Precipitation was further analysed but it did not show a straightforward relationship that contributes to the

GPP variation. The input vector data that include SWR, VPD, TEMP and TMAX showed a strong influence on the GPP behaviour. Table 5-3 shows the input variables sets that were analyzed.

Table 5-3 Input data sets.

Set	Input variables	Output variable
1	PRE, SWR, VPD, TEMP, TMAX, TMIN	
2	SWR, VPD, TEMP, TMAX	GPP
3	VPD, TEMP, TMAX	
4	VPD, TMAX	

Explicitly cross-validation in the LRM, MT and IBL was not performed; instead, generalization was ensured by the strategy of "early stopping of learning", i.e. deliberately undertraining the model and thus reducing accuracy on training set.

5.3.2 Comparison of DDMs

The input variable sets represent the combination of meteorological variables that affect the GPP variation. Notably, the input variables set 2 (which considers SWR, VPD, TEM and TMAX) stands out and show to be a good combination set of variables. Input variable set 3 slightly declines its performance emphasizing the sensitivity of SWR when it is not taken into consideration (Table 5-4). Major errors appear in the performance of ANN for both time frames. The final model structures are shown in Table 5-5.

Table 5-4 NRMSE of DDM (LRM, MT, IBL, ANN) for GPP simulation at the three altitudinal ranges on the testing data set at a bimonthly (15d), and monthly (30d) time frame.

Model	Input	R1		R2		R3	
		15d	30d	15d	30d	15d	30d
LRM	Set 1	0.294	0.270	0.272	0.245	0.460	0.437
	Set 2	0.292	0.272	0.274	0.242	0.466	0.447
	Set 3	0.289	0.271	0.268	0.266	0.483	0.476
	Set 4	0.293	0.271	0.266	0.252	0.483	0.476
MT	Set 1	0.316	0.270	0.278	0.251	0.467	0.437
	Set 2	0.292	0.272	0.279	0.242	0.473	0.447
	Set 3	0.281	0.253	0.280	0.233	0.483	0.476
	Set 4	0.293	0.253	0.262	0.252	0.483	0.476
IBL	Set 1	0.282	0.259	0.276	0.222	0.475	0.464
	Set 2	0.280	0.237	0.276	0.208	0.482	0.462
	Set 3	0.274	0.238	0.269	0.213	0.501	0.493
	Set 4	0.274	0.238	0.269	0.214	0.501	0.493
ANN	Set 1	0.335	0.441	0.410	0.472	0.504	0.534
	Set 2	0.356	0.399	0.327	0.419	0.509	0.498
	Set 3	0.326	0.424	0.318	0.461	0.555	0.569
	Set 4	0.304	0.336	0.293	0.305	0.476	0.470

Table 5-5 Model structures, number of linear model for MT, neighbours for IBL and hidden nodes for ANN.

Model	Input	R1	R2	R3
MT	Set 1	3	3	1
	Set 2	3	3	1
	Set 3	3	1	1
	Set 4	1	3	2
IBL	Set 1	15	9	18
	Set 2	19	20	7
	Set 3	19	18	26
	Set 4	16	18	26
ANN	Set 1	5	6	5
	Set 2	5	11	5
	Set 3	5	5	6
	Set 4	5	6	5

We analysed the performance of each DDM using the input variables data set 2 as the one leading to the smallest NRMSE. The Taylor diagrams in Figure 5-3, show the coefficient correlation, standard deviation and RMSE difference between the biomodel GPP and the surrogate data driven model. Performances of models for low and mid elevations are similar. IBL showed the capability to represent the GPP variation with minimum error, whereas ANN had the largest RMSE. For low and mid elevations (R1 & R2), IBL has the highest correlation coefficient for input variable Set 2 ($r = 0.80$ and 0.84, respectively) and lowest RMSE difference (RMSE = 0.037 and 0.026, respectively). On the contrary, ANN does not perform well showing correlation coefficient $r < 0.6$ and a large RMSE difference (RMSE > 0.045). For higher elevations (R3), none of the DDM showed good results in terms of correlation coefficient and RMSE.

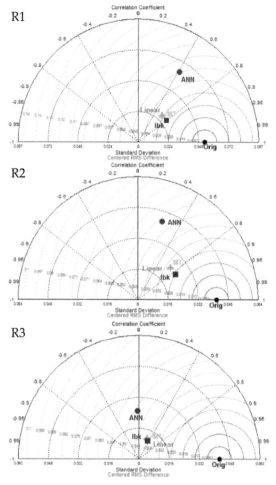

Figure 5-3 Taylor diagram showing pearson correlation, standard deviation and RMS difference between GPP testing set and each of the DDM technique (LRM, MT, IBL, ANN) for a monthly time frame using data Set 2 for all altitudinal ranges.

5.3.3 Order of effects

We compared the differences between the biomodel and surrogate model and assessed the advantage of using IBL using meteorological variables that change along an altitudinal gradient. The input variable set 2 was chosen as the best surrogate model to simulate GPP (smaller NRMSE). The final model structure is as follows:

$$GPP = f(SWR, VPD, TEMP, TMAX) \qquad (5\text{-}6)$$

5.3.3.1 Time frame

The IBL had better performance when data are aggregated in a monthly time frame. The prevailing pattern can be seen in Figure 5-4, in which the surrogate model tends to underestimate the phenomena. It can harmonize the peaks and dips during the growth and decay season of the vegetation.

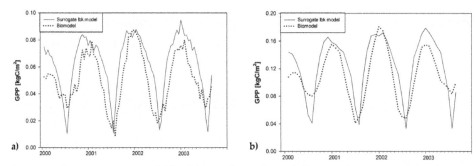

Figure 5-4 Bio vs surrogate model for GPP simulation for mid-elevation R2 (4200-4400 m a.s.l.) using IBL for bimonthly and monthly time frames

5.3.4 Computational time

The biomodel BIOME-BGC used to estimate the GPP variation at each elevation took approximately 40 seconds for a 12-year time series. To run the biomodel for the entire basin, we might need more than 4 days to run pixels of 30m x 30m. During the DDM analysis, less than 1 second was needed for IBL to run training and testing for a 12-year time series data for all altitudinal ranges. This shows the usefullness of having DDM-based surrogate of a process-based model for experimentation and "what-if" analysis.

5.4 Discussion

5.4.1 GPP responses to climatic variables

The model-free technique used to identify and select the appropriate input variables showed the strongest relationships between GPP and each of the meteorological variables. However, it is difficult to attribute the variation of GPP to a single meteorological variable since most of them are indirectly dependent of each other (Jung et al., 2007). We found it crucial to investigate the strength and relationship between each of the climatic variables with GPP for uncertainty analysis in future climate projections

5.4.2 Surrogate model performance

IBL showed capability to represent the GPP fluctuations at a monthly time frame and it seemed to be able to capture the signal trend of the seasonal dynamics.

The biomodel BIOME-BGC is a continuous model that updates state variables while the DDMs look for the connections between the system variables without taking into account the explicit knowledge of the physical behaviour of the relationship between GPP and the climatic variables. There are some inconsistencies in the DDM output (e.g. larger error differences the beginning of the time series and the overestimation of GPP during the decay seasonal dynamics), but this could be explained by the particulars of the training data set rather than limitations of DDMs. Other studies (Levin, 1998; Scardi, 1996) have demonstrated that GPP fluctuations can be estimated just to a certain extent by conventional linear models due to the complexity of this terrestrial ecosystems. It would appear that it is equally important to analyse the scope at which the meteorological conditions perform on the GPP variation or whether the GPP is a consequence of the system's self-organization. In this regard, IBL can be used as quite accurate simulators of GPP variations besides the presence of a short and noisy time series data set.

The weak relationship between the climatic variables and the GPP discovered for higher elevations (R3) can be perhaps explained by the ambiguous responses and uncertainty in the model forcing fields when vegetation is located in the upper line close to the permanent snow. An in-depth analysis of the climatic data sets and the joint effects of them on the variations of GPP are beyond the scope of this study. However, an improved reanalysis of the meteorological forcing and its interpretation would reduce uncertainties in future long-term time-series data. We already considered the orographic factor that we believe it may have a stronger influence due to study area located in a mountainous region in the Ecuadorian Andes.

5.4.3 Performance based on variables selected and time frame

5.4.3.1 Changes of Input variables set

When looking at the outputs from the different sets of input data, the results highlighted the important role of SWR, VPD and temperature as climatic forcing processes on the representation of GPP. These variables are related to the vapour flow and transpirational demand which influences the amount of moisture that the plant tissues are exchanging with the atmosphere and therefore the capture of CO_2 that is entering the stomata.

5.4.3.2 Changes of time frames

The temporal resolution does play an important role in the changes of GPP especially in these high-altitudinal ecosystems, since they are characterized by slow decomposition and growth rates and therefore slow processes which agrees with previous studies (Minaya et al., 2016; Spehn et al., 2006). This happens when the climatic variables are upscaled from daily to monthly values. It would appear that the GPP variation is not susceptible to sudden changes of high or low temperatures or presence of heavy rains at a regional scale.

5.4.4 Computational time

Current physically-based ecosystem models aim to create more realistic simulations but they ended up increasing the complexity and consuming lots of processing power. Additionally, determining appropriate values for the input data requires great diligence and for our case required an extensive fieldwork and analysis to get accurate values. Small uncertainties in the parameters may propagate to a wide range of variability in the simulations. As a follow up, simplifications are necessary but without compromising the capacity of the models. The surrogate model based on a data driven techniques was computationally faster and can easily be used to upscale and predict future climate scenarios from global climate data. In a simple exercise we compared the computational time used to estimate the GPP for one scenario of the Coupled Model Intercomparison Project (CMIP) models. In brief, to run a spatially distributed model of 10,300 cells using BIOME-BGC needs approximately 320 days in comparison to the almost 100 times faster IBL. The ecosystem process model requires a substantial investment of computational time in contrast to the DDM, which is shorter with enough accuracy for using it in multi-model runs.

5.5 Conclusions

This paper explored the use of surrogate DDMs to simulate the GPP along an altitudinal gradient in the páramos ecosystem in the Ecuadorian Andes. This was done by identifying the best input variables set for different time frames, for which the two methods were used - model free and model based. Based on the selected inputs, several models have been built (LM, MT, IBL, ANN), and for the considered use case, IBL model showed the best ability overall to reproduce the biomodel across a continuum of temporal scales. It was more responsive and sensitive to SWR, VPD, TEMP and TMAX as the main drivers for GPP on a monthly basis. However other models have also showed reasonable performance and all of them can be recommended for use in similar circumstances.

The IBL surrogate model does not replace a detailed and comprehensive physical based model, but it is a complementary statistical technique that link input and output variables for a comprehensive analysis of the ecosystem. The short computational time in running of the IBL will allow the extrapolation to higher temporal scales in future estimations of gross primary production (GPP), especially when using climate change scenarios such as the CMIP (Coupled model intercomparison project). Fast running surrogate models allow for experimentation and "what-if" analysis. In case of availability of more observations DDMs can be built not on the basis of data generated by the process model, but directly on measurements, and this paper confirms that it is possible.

Although the DDM techniques tested in this paper showed that precipitation was not a variable that influence the variation of GPP, it is well known that precipitation is the major driving force for plant growth and therefore carbon uptake by plants. The ability of the DDM techniques to model the climate scenarios and the sensitivity of GPP to precipitation deserves further studies due to the high number of complex biological processes (i.e. adaptation to climate, to nutrients availability and others).

Further research will focus on a more detailed analysis of how frequency, timing and amount of precipitation would influence thresholds for carbon uptake by the grassland and therefore GPP in the region, rather than using the total precipitation as investigated in this paper. Parameters such as leaf area index and available nitrogen content in soil are equally important and could be included within the input set of parameters. Since the photosynthesis is the main process for primary production which is also driven by indirect controls that operate environmental conditions on the mineralization of nitrogen due to litter and soil organic matter decomposition (Running et al., 2000).

Appendix 5-A

Description of the DDM used to evaluate the performance of BIOME-BGC model

M5 Model tree

The M5 (Model trees, version 5) is a piece wise linear model with submodels built based on a certain criteria that allow to do it in a more automated fashion (Quinlan, 1992). This splitting criterion is based on the standard deviation reduction (Witten and Frank, 2000) and the calculation of the expected reduction in error as a result of testing each parameter. The target is to determine which parameter is the adequate to split the portion of the data set that reaches a node and maximizes the expected error reduction, calculated as:

$$SDR = sd(T) - \sum_i \frac{|T_i|}{|T|} sd(T_i)$$

where T_1, T_2 ... are the subdata sets after the splitting of a chosen parameter. The splitting process finishes either when the standard deviation at that node is less than 5% of the standard deviation of the original data set (Solomatine et al., 2008) or when just a few instances remain. The instances are described by a fixed set of parameters and their values. The minimum number of instances in a leaf can be set to reduce overfitting problems. The advantage of MT is the capability of dealing with high dimensionality of parameters and its interpretability in solving physical problems (Solomatine and Dulal, 2003).

The Figure 5-5 shows the splitting of the input data followed by a leaf that has a linear model based on some other parameters. More details are given in Wang and Witten (1997).

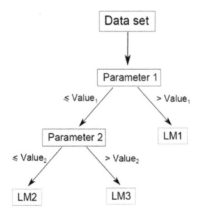

Figure 5-5 Model tree structure for data set, where Parameter 1, Parameter 2 represents the best attribute that maximized the expected error reduction at each node and LM1, LM2, LM3 are the linear regression models that follow $y = a_0 + a_1x_1 + a_2x_2$.

Instance-based learning

The instance-based learning (IBL) method applied here follow the conventional approach (Aha et al., 1991; Mitchell, 1997), unlike the other learning methods, it does not build a model first instead it stores the training values and delay the processing until a new instance must be classified. It does not use all the training data but rather some of the instances to build a local model. The k-nearest neighbour regression method (IBL) was used for this analysis, where the new instance x is the average of the outcomes of its k-nearest neighbors (usually k > 1) using a distance metric, which for this case was the standard Euclidean distance d.

$$d = \sqrt{\left(a_1^{(1)} - a_1^{(2)}\right)^2 + \left(a_2^{(1)} - a_2^{(2)}\right)^2 + \cdots \left(a_k^{(1)} - a_k^{(2)}\right)^2}$$

The mean nearest neighbour distance is calculated as follows:

$$\overline{d_{Ibk}} = \frac{\sum_{i=1}^{n} d_i}{N}$$

where N is the number of points and d_i is the nearest neighbor distance for the point i

As an example Figure 5-6 shows a IBL regression method using different number of nearest neighbours.

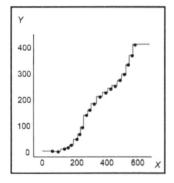 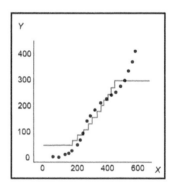

Figure 5-6 The nearest neighbour regression method using *k* nearest neighbours when a) k=1, and b) k= 9

Artificial Neural Networks

The artificial neural network (ANN) model most widely used is the multi-layer perceptron, that represents the non-linear complexity of a region of the input space.

In general an ANN can be graphically represented by a number of interconnected nodes arranged in three layers: input, hidden nodes and output (Figure 5-7) and in most of the cases contains an additional node with a constant value of 1, which is added in the input and hidden nodes layers to contemplate the bias term. The ANN requires a training process to be able to simulate input output relationships. In general this is done by adapting the weights, which are the strength of the connections by an optimization algorithm. The main algorithm used is the gradient based, which looks at the steepest relation between performances measured in the output space. The "learning" finishes when the error measured is lower than a threshold or a number of specific iterations is reached. More details are given in Rumehalt et al. (1986), who develop the error propagation algorithm and improvements in later studies done by Li et al (2012) and Kasabov (1996). The implementation used to optimize the ANN in this research followed the gradient descent algorithm with regularization available in the MATLAB NN toolbox (Hudson et al., 2014).

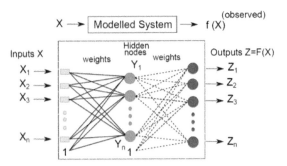

Figure 5-7 ANN (multi-layer perceptron) structure, where X_1, X_2,...X_n are the input parameters, Y_1,...Y_n are the hidden nodes, Z_1,...Z_n are the outputs and the lines that connect the layers are the weights. Inputs and output should be transformed [0,1].

Conceptualization catchment processes it's simply too complex
(Tetzlaff et al.)

6

RUNOFF GENERATION IN A COMBINED GLACIER - PÁRAMO CATCHMENT

Hydrological processes in combined glacier and *páramo* catchments are of great interest especially for the surrounding population that receive most of their consumptive water from these sources. Previous studies have shown that the melting of glaciers contributes to runoff generation and that the *páramo* ecosystem acts as a natural sponge, which plays an important role in regulating the runoff during dry-season. However, not all runoff processes are well understood in the Andean Region due to the high spatial variability of precipitation, young volcanic ash soil properties, soil moisture dynamics and other local factors such as vegetation interception and high radiation that might influence the hydrological behaviour. In addition, there is a lack of evidence of the origin and quantification of the contribution of runoff components in the *páramo* ecosystem. This chapter focuses on data collection and experimental investigations in the particular case study catchment. The approach consists of the identification of suitable environmental tracers and hydrochemical features to identify the various runoff sources in order to determine their respective contribution during dry and wet conditions.

This chapter is based on:

Minaya, V., Camacho, V., Weninger, J., and Mynett, A. E.: Quantification of runoff generation from a combined glacier and *páramo* catchment within an Ecological Reserve in the Ecuadorian highlands, doi:10.5194/hess-2016-569, *Manuscript under review for Journal Hydrology and Earth System Sciences*, 2016.

6.1 Introduction

Páramos are a important source of water for agriculture and urban use (Buytaert and Beven, 2011) and sustain a great biodiversity and unique ecological processes (Hofstede et al., 2002; Madriñán et al., 2013). The importance of the *páramos* is associated with the tremendous capacity of water retention in its volcanic ash soil covered by vegetation (Poulenard et al., 2002; Roa-García et al., 2011). Ecuador has nearly 12500 km² of *páramo* of which 64% in the area above 3000 m a.s.l. has been transformed or degraded (Hofstede et al., 2002) and the remaining areas are currently under constant pressure. At a higher altitude (>4000 m a.s.l.) these ecosystems are influenced by permanent snow and glaciers that feed directly the river's drainage system or might contribute further downstream by resurgence (Cauvy-Fraunié et al., 2013; Favier et al., 2008; Villacis, 2008).

A lot of attention has been put on the relationship between climate change and the retreating of glaciers globally (Beniston, 2003); specially because climate change in conjunction with the rapid change in land use can jeopardize the water quantity and quality of the *páramos* (Buytaert and Beven, 2009; Buytaert et al., 2006a; Jansky et al., 2002). Bradley et al. (2006) showed a clear evidence that the surface temperature changed faster in higher compared to lower elevations in the Tropical Andes with a rate of 0.11°C per decade in the period 1939 to 1998. The concern of the scientific community lies in the implications of the melting water and their contribution to the hydrological systems beside the response of the terrestrial, aquatic biota (Cauvy-Fraunié et al., 2013) and water security for communities that rely on these catchments in the tropical regions (Brown et al., 2010; Kaser et al., 2010). The complexity of these glacierized-*páramo* catchments increases since they are more affected than those in the temperate regions due to a continuous ablation at all altitudes (Kaser and Osmaston, 2002) and show as a consequence a change in the hydrological, geomorphic and ecological processes. Due to the orographic properties of these high mountainous regions in the *páramos*, the precipitation regime has a remarkably large spatial variability (Buytaert et al., 2006b; Celleri and Feyen, 2009).

Several studies enhance the importance of a fair understanding of the hydrological complexity of these interconnected systems and the implications for water resources management in the region (Buytaert and Beven, 2011; Buytaert et al., 2010; Carrillo-Rojas et al., 2016; Cuesta et al., 2013; Viviroli et al., 2011). Tracer experiment analyses have been used widely to provide more information about the connectivity and time scales of the contribution of the main runoff sources and flow pathways (Condom et al., 2012; Dahlke et al., 2012; Huss et al., 2008; Munyaneza et al., 2012; Villacis et al., 2008; Wenninger et al., 2008; Windhorst et al., 2013). However, a suitable spatial hydrochemical characterization and quantification of the groundwater reservoirs and meltwater infiltrations in these catchments of complex geology and topography remains a challenge. In particular, in the hydrological subsurface processes such as percolation, lateral subsurface flow, and groundwater recharge (Hindshaw et al., 2011; Nelson et al., 2011).

In this regard, this study aims to provide effective tools to investigate the origin of the main runoff components and to quantify their contributions using environmental tracers (isotopes and major ions). In addition, this study comprises a complete hydrochemical analysis of the runoff components separated by source and location. This will provide a fair understanding of the hydrological interactions of a glacierized-*páramo* system.

6.2 Materials and methods

6.2.1 Study area

The Los Crespos - Humboldt catchment is described in Section 2.2 and its main climatological and geological characteristics from Sections 2.4 to 2.6, respectively.

6.2.2 Data collection

The catchment is equipped with 2 hydro-meteorological stations as described in Section 2.5 (Figure 2-6). The station located at the outlet of the catchment is the Humboldt station (4010 m a.s.l.), which records data of temperature and electrical conductivity of the water. Isotopic and hydrochemical samples were collected in July 2014. Figure 6-1 shows the catchment with the sampling sites. We assigned IDs to the sub-catchments based on the Pfafstetter coding system (Pfafstetter, 1989) that determine upstream-downstream relationships between sub-catchments based on the code alone. The advantage of using this topological encoding system is that it efficiently identifies the main river system and continues with the delineation of subcatchments based on the convention of increasing ordinal values. The catchment contains two main tributaries that originate from the Antisana glacier, the difference is that the first one flows through large boulders and rocks of different size until it meets the second one that flows through *páramo* vegetation. The highest monthly average flow during the year occurs in June, with an average of 300 l s^{-1}; whereas the lowest monthly flow occurs in March with an average of 200 l s^{-1}. Due to the technical and logistics limitations in the study area, flow was measured only at the outlet of the catchment, and precipitation was assumed to be the same in the entire catchment.

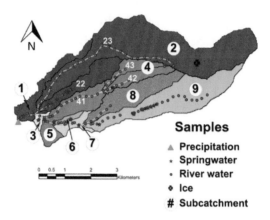

Figure 6-1 Investigation area with sampling sites for precipitation, spring water, surface water and ice.

6.2.3 Experimental set-up

6.2.3.1 For dry conditions

We refer to dry conditions when precipitation was absent for at least 3 consecutive days. Water samples were taken from the main streams including all tributaries (n = 107), springs (n = 44), ice (n = 3) as shown in Figure 6-1 during dry conditions. Every 200 m, electrical conductivity (EC) and temperature (°C) were measured in-situ using a WTW LF340 series conductivity meter. A volume of 2 ml was collected in a glass bottle (PTFE/silicone septa) filled to the top for stable isotopic analysis ($\delta^{18}O$, δ^2H) to prevent evaporation. In addition, every 400 m two polyethylene bottles of 25 ml water samples were filtered for analysis of major anions (Cl^-, SO_4^{2-}) and cations (Ca^{2+}, Mg^{2+}, Na^+, K^+). The latter bottle was previously prepared for preservation by adding a drop of nitric acid (HNO_3). All vials were kept in a cooling box at 1 - 5°C.

Additional samples were also analyzed for SiO_2 using the 8185 method of Silicomolybdate using a Hach DR890 Portable Spectrophotometer and for HCO_3^- using a Hach Digital Titrator. Additional information such as weather conditions, GPS coordinates and a description of the sampling site was also recorded. Figure 6-2 depicts the variety of the sampling sites used within the catchment.

Figure 6-2 Sampling sites a) Surface water flow within the *páramo* vegetation, b) Surface runoff from the glacier, c) Junction of two main streams (from glacier and *páramo*), d) swampy areas mixture of surface and subsurface runoff, e) spring water and f) lake located in the moraine just after the melting of the glacier. The red arrow indicates the flow direction.

6.2.3.2 For wet conditions

During rainfall events, the sampling of the surface water and precipitation was done only at the Humboldt Station for rainfall-runoff analyses. Samples were collected with a resolution of 15-20 min during events.

Rainfall samples were collected using a self-made device that consists of a funnel (diameter 140 mm) with a micro filter at the neck level joined to a drip chamber and to a 60 cm PVC tubing connected at the end to a rigid needle (3.8 cm, 28G) embedded in rubber cork of a polyethylene vacuum packed bottle (1L) (Figure 6-3). We use one device per rainfall event and the sample was immediately extracted after or during the event following the procedure of the IAEA (IAEA/GNIP, 2014).

All samples were analyzed for major anions, cations and isotopes using the same procedure as described above in section 6.2.3.1 and SiO_2, EC and temperature was measured in-situ.

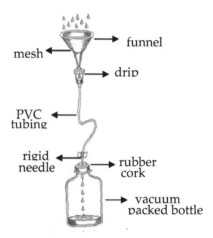

Figure 6-3 Self-made rainfall collector.

6.2.4 Laboratory Methods

Oxygen and hydrogen isotopic values, expressed in ‰ in relation to the Vienna Standard Mean Ocean Water (VSMOW), were measured using the Liquid-Water Isotope Analyzer (LGR DLT-100, precision <0.3‰ for $^{18}O/^{16}O$ and <1.0‰ for $^2H/^1H$).

The samples for major cations were determined by mass spectrometry using the Thermo Fisher Scientific XSeries 2 ICP-MS (limit of quantification ~2 ppb). Anions were analyzed by using ion chromatograph Dionex ICS-1000 (limit of quantification 2000 ppb). All analyses were performed following quality assurance and control procedures of the laboratory at UNESCO-IHE.

6.2.5 Data analysis

6.2.5.1 Spatial hydrochemical characterization

One-way ANOVA tests were used for significant differences ($P < 0.05$) for the different runoff sources e.g. ice, precipitation, surface water and spring water (shallow subsurface flow). The same test was applied for a location analysis related to subcatchments and geological background. If the results showed a large variation at subcatchment level, those were again divided in smaller subcatchments for a detailed analysis to identify the cause. Significant t-tests were followed by Tukey multiple comparisons as post-hoc tests indicated by a lowercase superscripted letters on top of each boxplot.

In addition, to assess the spatial distribution of hydrochemical parameters in the catchment, concentrations were plotted against distance from the outlet.

6.2.5.2 Flow pathways and routing

Tracers enabled the identification not only of the runoff sources but also the quantity that contributes to the river flow. We applied the mass-balance approach from downstream to upstream to calculate the discharge when two streams met at the confluence point (Eq. 6-1 and Eq. 6-2)

$$Q_T = Q_1 + Q_2 \tag{6-1}$$

$$C_T Q_T = C_1 Q_1 + C_2 Q_2 \tag{6-2}$$

where Q_T is the total runoff, Q_1, Q_2 are the runoff components in m³/s and C_T, C_1, and C_2 are the concentrations of total runoff, and of runoff components in mg/l or ‰.

6.2.5.3 End Member Mixing Analysis (EMMA) and Hydrograph Separation

A principal component analysis (PCA) based on the method described by Christophersen & Hooper (1992) was carried out using the water quality parameters obtained. Mixing diagrams of EC (µS/cm), SiO_2 (mg/l), Cl⁻ (in mg/l) and δ^2H and $\delta^{18}O$ (‰VSMOW) indicated their suitability as tracers for the hydrograph separation.

Isotope and hydrochemical data were combined with discharge data to perform three component hydrograph separations based on steady state mass balance equations and hydrograph separation assumptions (Buttle, 1994; Pearce et al., 1986; Uhlenbrook et al., 2002). We included a third runoff component in Equations 6-1 and 6-2 to calculate three component hydrograph separations for the total runoff (Q_T) (Eq. 6-3).

$$C_T Q_T = C_1 Q_1 + C_2 Q_2 + C_3 Q_3 \tag{6-3}$$

Rainfall characteristics, including duration, total rain, maximum and average intensity were estimated for the rain events. A rainfall event was defined as a rainfall occurrence with rainfall intensity greater than 1 mm/hr, and intermittence less than four hours. Peak flow, water depth, and time to peak were determined for each event.

Analytical and tracer end-member uncertainties were accounted for the hydrograph separation and quantification of the runoff components based on a Gaussian error propagation technique with 70% confidence interval (Eq. 6-4) (Genereux, 1998).

$$W = \left\{\left[\frac{\partial y}{\partial x_1} W x_1\right]^2 + \left[\frac{\partial y}{\partial x_2} W x_2\right]^2 + \cdots + \left[\frac{\partial y}{\partial x_n} W x_s\right]^2\right\}^{\frac{1}{2}} \qquad (6\text{-}4)$$

, where W is the uncertainty of the each runoff component in %, $W x_1 x_2$ are the standard deviations of each end-member, $W x_s$ is the analytical uncertainty and $\frac{\partial y}{\partial x}$ are the uncertainties of the runoff component average contribution regarding the tracer concentrations.

6.3 Results

6.3.1 Hydrochemical catchment characterization

6.3.1.1 Runoff sources

We clustered all samples in four major groups based on the runoff sources: Ice (n = 3), Precipitation (n = 4), Surface water (n = 107) and Spring water (n = 44) and compared the groups for significant differences. SiO_2, K^+ and stable isotopes (δ^2H and $\delta^{18}O$) gave a first glimpse of the composition of these groups by showing significant differences among all groups (Figure 6-4). However, most of the tracers showed a high variability within each group due to the mixture of contrasting backgrounds of water samples. Major ions and EC values in Ice and Precipitation showed lower values in comparison to the other groups, which require a distinctive classification in order to give more informative results. For this reason, surface and spring water samples were disaggregated and grouped in subcatchment and geological background, respectively.

Surface runoff – Subcatchment analysis

In this second approach only samples within the Surface water category were classified in catchment and subcatchment groups to give more confidence intervals in the results (refer to Figure 6-1).

The major ions and EC values from the subcatchments 41, 42, 43, 6, 7, 8 and 9 did not show any clear contrasting pattern among them thus can be considered as a single group (Figure 6-5). These subcatchments belong to catchments 4 and 5 which lie in a highly vegetated side of the catchment. Whereas subcatchment 21, 22 and 23 that belong to catchment 2 were significantly different ($p \leq 0.05$) and therefore should remain grouped separately (Figure 6-5). Catchments 4 and 5 from now on will be referred to as *páramo* catchment. Catchment 2 lies on a mixed catchment of water coming directly from the glacier (subcatchment 23) without any other type of contribution and water from a small tributary of a combined source of surface and spring water (subcatchment 22). Subcatchment 23 from now on will be referred to as glacier catchment.

All major cations showed significantly lower values of concentrations for subcatchment 23, thus demonstrating a unique hydrochemical characteristic of streams derived from glacier components.

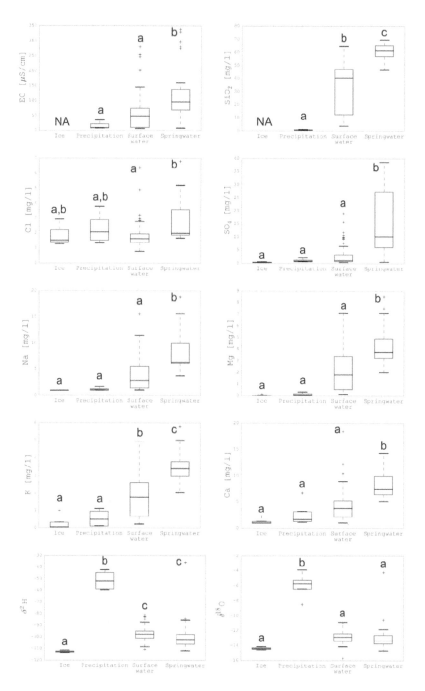

Figure 6-4 Chemical components and stable isotopes of water samples within the Los Crespos-Humboldt basin, analyzed per runoff source (Ice, Precipitation, Surface and Spring water). Lowercase letters indicate significant differences among sources ($P \le 0.05$), according to Tukey's test. NA= samples are not available.

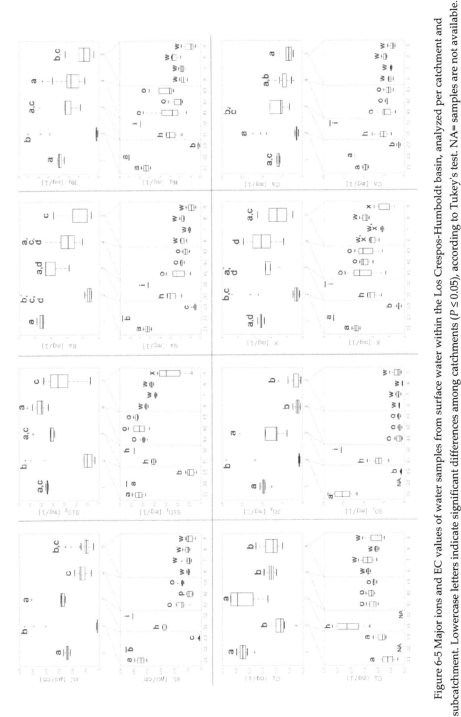

Figure 6-5 Major ions and EC values of water samples from surface water within the Los Crespos-Humboldt basin, analyzed per catchment and subcatchment. Lowercase letters indicate significant differences among catchments ($P \leq 0.05$), according to Tukey's test. NA= samples are not available.

Surface runoff – Flow paths and routing

Each of the previous analysis was important for the spatial hydrochemical characterization of surface and spring waters within the Los Crespos-Humboldt catchment. A PCA was carried out to identify interrelationships between major ions, we selected a smaller set: EC, SiO_2, Na^+, K^+, $\delta^{18}O$, and δ^2H for further analysis.

Distance to the outlet and altitude were highly correlated and assumed to have a linear response despite the weather conditions which can differ at different locations within this high altitudinal mountain ecosystem. We selected distance to the outlet to display the effect of surface water within the subcatchment and the confluence with other tributaries along the way to the outlet (Figure 6-6). The major ions and EC values showed a significant spatial variability and evidently separate two contrasting groups: surface water directly coming from the melting of the glacier (subcatchment 23) and the surface water that comes from the vegetated areas (subcatchments 41, 42, 43, 5, 6, 7, 8, 9).

It should be noted that subcatchment 9 starts also with a small contribution of glacier but most of the water comes from the *páramo* vegetated areas. The high concentration values of subcatchments 22 and 31 are due to a considerable amount of spring water contribution to the main channel as well as the small variations displayed among samples within the same subcatchments.

Spring water

Spring water characteristics were based on the geological formation from which they originate (refer to Figure 2-7). Spring water samples that come from the Lahar Rojo formation were significantly higher in most of the major ions and EC concentrations (Figure 6-7). The rest of the geological formations showed different concentration ranges; however, they could not be tested for significance due to the lack of samples for specific major cations.

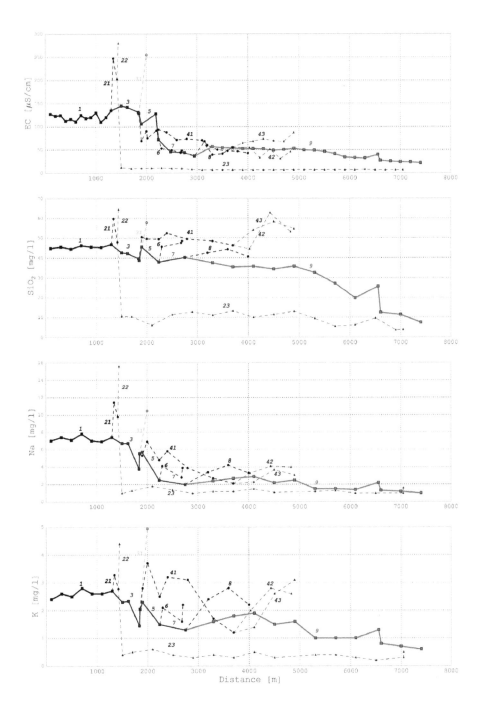

Figure 6-6 Chemical components of surface water samples within the Los Crespos-Humboldt basin, analyzed by distance to the outlet, numbers indicate subcatchment group. The thick solid line represents the main stream and dashed lines are the tributaries.

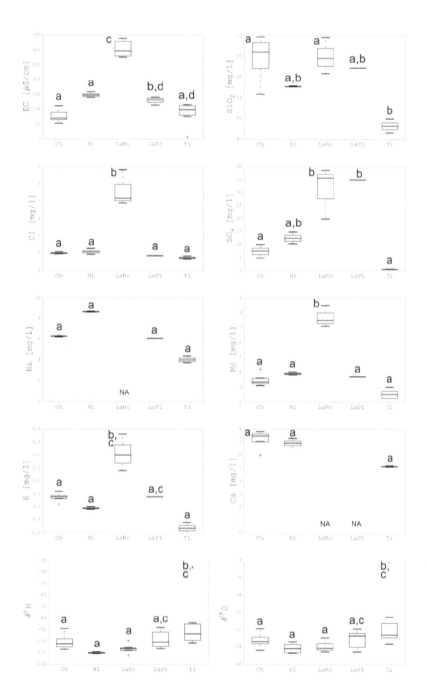

Figure 6-7 Chemical components and stable isotopes of spring water samples within the Los Crespos-Humboldt basin, analyzed per geological background (Ch = Chacana volcanic rocks, Hi = Hialina lava, LaRo = Lahar Rojo, LaPl = Lavas Pleistocene, Ti = Tillite late ice age). Lowercase letters indicate significant differences among geological background ($P \leq 0.05$), according to Tukey's test.

6.3.2 Rainfall events

The sampling during rainfall events corresponds to low-medium intensity rain (2 mm total rain) and one of them was considered as representative and analyzed for rainfall-runoff evaluation. The duration of the event was 12 hours with a maximum intensity of 0.3 mm/h and an average intensity of 0.18 mm/h.

Isotope composition for all samples in the catchment is shown in Figure 6-8. Precipitation ranged from -8.5 to -3.9 ‰ for $\delta^{18}O$ and from -59.5 to -42.4 ‰ for δ^2H. Ice samples have a lighter isotopic signature, the lightest value corresponds to a 30 cm-depth sample, followed by a 20 cm-depth sample and the slightly enriched value belongs to the 10 cm-depth sample. We clustered the samples in such a way that we clearly identify the isotopic signature of the main river and tributaries that come from the *páramo* component as well as the glacier component. The signature of the *páramo* component shows a relatively lighter or more depleted value of isotopes in comparison to the signature of the glacier component, which are heavier or enriched. Spring water signature showed a wide range of isotopic composition from -14.8 to -10.6 ‰ for $\delta^{18}O$ and from -112 to -84.1 ‰ for δ^2H. Likewise, the isotopic composition of storm runoff ranges from -13.03 to -11.2 ‰ for $\delta^{18}O$ and from -99 to -89.3 ‰ for δ^2H. The storm runoff samples are the ones collected in the main stream at the outlet of the catchment and include samples of pre and post event, which will be identified in the mixing plot analysis (Figure 6-9).

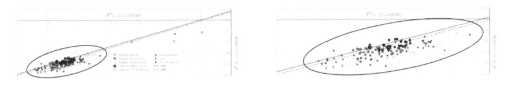

Figure 6-8 Stable isotope compositions of precipitation, surface water, springs, ice, storm runoff. GMWL: $\partial^2H = 8.13 \times \partial^{18}O + 10.8$ ‰. (Source: Rozanski et al. (1993)). LMWL for Izobamba: $\partial^2H = 8.1 \times \partial^{18}O + 12.8$ ‰ (Source: IAEA (2016)).

6.3.2.1 Mixing plots

We derived mixing plots with all possible combinations of the small set of parameters (EC, SiO_2, Na^+, K^+, $\delta^{18}O$, and δ^2H). We considered three main components: precipitation, glacier and *páramo*, each of which has its own chemical signature and serves as a vertex of a triangle that defines the boundaries of the storm runoff concentrations.   EC and stable isotopes ($\delta^{18}O$, and δ^2H) were identified as conservative tracers that characterized the end-member concentrations represented by precipitation, glacier and *páramo* runoff (Figure 6-9). Most concentration points of the storm water were located between the runoff from *páramo* and glacier. The mixing plot shows the evolution of the stream water before (pre), during (event) and after (post) the event. The discharge started with high EC values of around 130 µs/cm, decreasing to an average of 70 µs/cm during the event and rising up to 95 µs/cm after 12 hours of the event. During the event the storm water is showing slightly heavier isotope values.

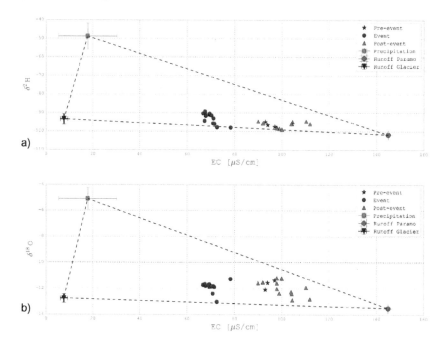

Figure 6-9 Mixing diagram showing stream water evolution and end-member EC and stable isotopes: a) δ^2H, and b) $\delta^{18}O$ during the July 2014 rainstorm.

6.3.2.2 Hydrograph separation

Dry conditions

An initial estimation of the flow percentages during rainless periods was derived with EC, stable isotopes, and major ions independently. The flow path estimations for major ions were not considered because the concentrations and/or differences were too low thus giving unrealistic estimations for the subcatchments. Thus, we used EC values and stable isotopes to estimate the contribution of water from the glacier component, which was of 21% for EC, 14% for δ^2H and 15% for $\delta^{18}O$. Likewise, the percentages of flow that come from the *páramo* vegetated areas are 52%

for EC, 71% for δ^2H and 78% for δ^{18}O, the remaining percentages come from small streams that join the main stream close to the outlet.

Wet conditions

A three-component hydrograph separation based on EC and δ^2H concentrations quantified the relative contribution of precipitation, glacier and *páramo* to the total flow. During the event, the total discharge was composed of 8% precipitation, 41% flow from glacier and 51% flow from the *páramo* component (Figure 6-10). The glacier component was the first to rise; both components have the maximum contribution during the peak time of the discharge. The rising limb mainly comprised similar contributions from glacier and *páramo* components and to a smaller extent by precipitation, whilst during the recession limb the contribution of precipitation increased. After the storm runoff there was still contribution from the *páramo* component while the contribution from the glacier decreased gradually as shown in Figure 6-10.

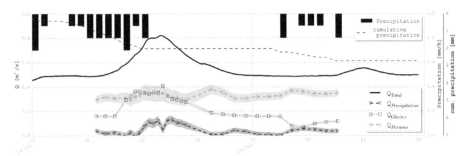

Figure 6-10 Three-component hydrograph separation, contribution of Glacier, Páramo and precipitation to stream runoff based on the EMMA using EC and δ^2H as tracers for 14 July rainstorm event. Colored area shows the estimated error propagation of the components.

6.4 Discussion

6.4.1 Spatial hydrochemical analysis and suitable tracers

Silica (SiO_2) and EC are highly correlated with weathering processes. They proved to be appropriate tracers to separate the main groups of runoff sources (precipitation, surface and spring water); whereas all major ions succeed only to significantly distinguish the spring water from the rest of the sources. In all cases, the spring water displayed higher values with large variations; this suggests that some runoff could originate from deeper sources, from fissures and fractures in the rock, where longer residence and contact times have increased the ion concentrations. Similar evidence was observed in other headwater catchments (Hugenschmidt et al., 2014; Scanlon et al., 2001). However, we should also pay attention to the outliers, which are a clear evidence of resurgence of the meltwater from the glacier consistently characterized with low values of EC. The latter confirms the meltwater infiltration during a study of the influence of the glacier on the water levels in the same catchment (Cauvy-Fraunié et al., 2013). Unfortunately, it is very difficult to crosscheck with the water signatures from the isotopes and water chemistry since they probably could be altered as a consequence of the strong influence of bedrock substrates, altitude and manifold underground processes (Nelson et al., 2011).

In a more comprehensive analysis, most of the spring water samples showed silica concentrations of 55 to 70 mg/l; while the samples that correspond to the type of geology Tillita showed concentrations between 45 and 50 mg/l. In most of the cases the latter type of geology consists of impermeable tough layers that have a shallow water table (Cuesta et al., 2013) and thus could easily get in contact with the subsurface flow and experience a dilution effect. The comparatively higher concentrations in the cations (Ca^{2+}, Mg^{2+}, Na^+, K^+) in spring water showed similar characteristics to silica. Hence these cations should be considered as indicators for water from deeper soil layers in study areas comparable to this one. These results strengthen the assumption that most of the spring water comes from a groundwater source; nevertheless the discrimination among groundwater, shallow and deeper subsurface flows are subjects of further analysis which lies outside the scope of this study.

The analysis of surface water per subcatchment identified two distinctive groups; the samples from subcatchments 4 and 5 were pooled together and defined as the *páramo* component. In the same way samples from subcatchment 23 were defined as the Glacier component since that was the stream that exclusively carried water from the melting of the glacier without any input of additional sources. However, during dry conditions in the diurnal cycle the contribution of the glacier might: a) increase as a consequence of the melting of the glacier due to the strong shortwave radiation especially in the low part of the glacier and b) decrease as consequence of sublimation of the glacier due to high wind velocities (Favier et al., 2004).

EC, silica and cations were appropriate tracers that displayed the significant difference between the glacier and the *páramo* component. The landscape played an important role in the influence of the hydrochemical composition of the surface water which was confirmed with the analysis of its distance to the outlet. The latter revealed the contribution of the spring water to the stream by showing changes in the concentration of the different tracers along each subcatchment and the shift in concentration while joining other subcatchments.

The stable isotopes ($\delta^{18}O$ and δ^2H) proved to be good tracers to distinguish between precipitation and ice samples; however they were unable to clearly distinguish between surface and spring water sources. This could be partially related to the fact that surface water samples were taken along the streams and encompass a combination of surface and spring water. Large variations in the stable isotopes could be also associated with the undetermined quantification of shallow and deep subsurface flow.

6.4.2 Quantifying the contribution to storm runoff

The separation of the different components and the contribution of each were challenging to determine and may change in an unknown amount due to unidentified layers and/or fissured and fractured rocks from which spring water originates. Usually, small head watersheds are a result of a mixing between rainfall, soil water and groundwater (Marechal et al., 2013). Nevertheless, since the water sampled in the rivers took into account the contribution of spring water, our main objective remains in the quantification of the relative contribution from glacier and *páramo* and their evolution with time as the main components for the total runoff at the outlet of the catchment.

The isotopic composition of rainfall and their relative distance to the GWML propose a possible evaporation hypothesis. Spring water signature has a wide range; the ones that showed heavier isotopic composition could be associated to a precipitation, wetland or shallow subsurface recharge; while the ones with depleted compositions could be linked to deeper subsurface layers considering also the high EC concentrations thus implying that the water could have been stored for a longer period of time. During the event the storm water contained more heavy isotopes which came from the rain or could be associated with wetlands/open waters or shallow subsurface flow (from previous rainfall events) and also with contributions from the saturated zone, which can be highly dynamic. The challenge to separate the runoff components in this catchment was investigated in earlier studies (Mena, 2010) that estimated an average contribution of 45% of the glacier component, just slightly above the estimation of 41% reported in the present study based on the EMMA using EC and δ^2H as tracers. Whereas for natural *páramo* it can vary around 50 to 70%, while this study reported around 51%. The remaining contribution from precipitation (approximately 8%) can be attributed to direct superficial runoff.

A representative end-member mixing analysis was carried out with three main components: precipitation, glacier and *páramo* as justified earlier. Some of the monitored stormflow samples were not fully confined within the triangles, which might increase the uncertainty in the evaluated event.

In order to overcome the limitations of non-conservative behavior of major ions, we preferred the application of stable isotopes for the hydrograph separation, which has been commonly used in tropical and subtropical areas (Elsenbeer, 2001; Goller et al., 2005; Klaus and McDonnell, 2013; Mul et al., 2008). For our case, a plausible approach was the combination of EC and δ^2H that demonstrated a clear separation of the three distinctive precipitation, glacier and *páramo* components likewise stated by Mena (2010). It is important to realize that these types of tropical Andean catchments have a high spatial variability of precipitation (Buytaert and Beven, 2011; Buytaert et al., 2006b; Celleri et al., 2007) due to orographic effects, which at a certain extent can influence and might mislead the quantification of the contribution of this component.

Based on the hydrograph separation, the contribution of the glacier component during the storm increases at a faster rate; this is mainly attributed to the fact that there is no water retention at any point in that subcatchment and yet the riparian area consists mainly of boulders, rocks and large soil particles that drain rapidly with a very low capability of holding moisture. Conversely, in the *páramo* component there are several zones with less steep slopes that are hydrologically disconnected due to the irregular terrain (Buytaert and Beven, 2011). And yet, they behave as floodplains, swamps and wetlands that dissipate the stream energy and buffer the peak flow at the outlet, contrary to what was found in a similar study by Buytaert *et al.* (2010). The soils of the riparian zone in the *páramo* subcatchments comprise of smaller soil particles that are poor in percolation thus offer a high water-holding capacity (Minaya et al., 2015a) and consequently a high water attenuation (Buytaert and Beven, 2011). Equally important is the interception of rainfall, which is the first process in a rainfall-runoff event. This interception particularly in the tussock vegetation should not be neglected and might contribute to a longer lag time during rainfall events and an increased recharge of water into the soil in these high altitude ecosystems (Buytaert and Beven, 2009; Janeau et al., 2015).

6.5 Conclusions

Estimates for the relative contributions of the main runoff components provided valuable information on the origin of the water various sources and hydrological characteristics as well as hydrochemical composition of the water cycle in this high mountainous region. Although it is not sure whether this runoff ratio will be maintained in future because it is linked to future climate drivers; we can certainly assure that a desirable plan for adequate water resources management should enhance the protection of the *páramos* as reservoirs of water in the highlands since they are the main contributors to runoff generation.

Although we only monitored rainfall events with medium intensity, the runoff patterns are in line with the expected dynamics within this catchment. The high contribution of the glacier component during rain events remains valid. We identified a couple of sources with clear evidence of resurgence of meltwater from the glacier, consistently characterized by low values of EC. It should be noted that during rainless events, there might be variability due to the diurnal cycle and contribution of melt water due to the exposure to solar radiation. Therefore, the effect of temporal resolution needs to be further studied since these streams depend on the glacier influence. Certainly, long-term analysis will contribute to a better understanding of the dependency of runoff generation on soil moisture and vegetation interaction.

The present study focused on a spatial representation of the main runoff components. However, there is room for improvement specifically in the correct separation of the shallow and deep subsurface flows as well as in the groundwater movement that for now was clustered into a single group as 'spring water'. The lack of soil moisture measurements and assumptions on the permanently saturated zones also add uncertainty to the quantification of the subsurface processes that regulate the contribution of surface runoff. Despite these limitations and uncertainties, the combination of stable isotopes and geochemical tracers was seem to improve the understanding of runoff processes in this combined glacier and *páramo* catchment in the Ecuadorian Andean Region, for which no runoff investigations were available before.

Acknowledgements

I would like to express our sincere gratitude to Fred Kruis, Ferdi Battes and Berend Lolkema for their assistance in the laboratory analysis.

Impossible is a word reserved for those without imagination
(A. Jayashankar)

7

A PROCESS-ORIENTED HYDROLOGICAL REPRESENTATION OF A PÁRAMO CATCHMENT

In Chapter 4, the results of the biogeochemical BIOME-BGC model did not provide enough understanding of the hydrological processes within the soil compartment. The aim of this chapter is to have a better representation of the soil hydrology through the use of a conceptual hydrological model in order to understand the influences of the vegetation and the soil properties in the hydrological system of the Antisana *páramo*. The conceptualization of the hydrological processes together with experimental information of fine-resolution data on bio-physical features was performed. The model contains a process-realistic description of the runoff generation mechanisms represented by several hydrological units which use linear and non-linear reservoir routines for spatially delineated cluster areas with the same dominating runoff generation processes. These hydrological units were established based on available knowledge and field experiments carried out and analysed in Chapter 6. The process-oriented hydrological Tracer Aided Catchment model distributed TAC^D was used to check if it is possible to determine the runoff generation processes in a *páramo* catchment. TAC^D seeks to reproduce water fluxes in a process-oriented way, both spatially and temporally distributed (30 x 30 m grid, daily mode) using the PCRcalc framework for dynamic modelling.

This chapter is based on:

Minaya V., Jianning, R., Mishra, P., Corzo, G., van der Kwast, J., Uhlenbrook, S., and Mynett, A. E.: Process-oriented hydrological representation of a *páramo* catchment; Ecuadorian Andean Region, Manuscript in preparation, 2016.

7.1 Introduction

Hydrological processes and particularly runoff generation processes are one of the most challenging issues in hydrological modeling in high-altitude catchments (Ott, 2002; Uhlenbrook et al., 2004; Buytaert & Beven, 2011), where some of these processes are determined by fractured aquifers in subsurface layers. Some studies have evaluated the performance of hydrological modeling in tropical regions in the Andes; however, simulations of water fluxes using physical equations have failed in areas of high heterogeneity and limited knowledge of physical properties (Wissmeier, 2005; Beven, 2006b; Savanije, 2008). Many hydrological processes in these hydrological heterogeneous ecosystems are far from being well understood (Buytaert et al., 2006b; Buytaert et al., 2010; Buytaert & Beven, 2011). The importance of these tropical catchments in the Andes lies in an adequate description of the hydrological processes including a good estimation of the storages for a fair quantification of water availability within the catchment (Winsemius et al., 2006). This is the case of Los Crespos-Humboldt catchment located in the Antisana volcano in the Ecuadorian Andean grasslands known as páramos. This area has become a benchmark for other climate change studies due to its geographical location, elevation, biodiversity and interest in the availability of water for the capital city of Ecuador. This catchment discharges in La Mica reservoir that produces hydropower and water supply for more than half-million inhabitants in the capital city of Ecuador and therefore the runoff generation processes and storage capacity are vital to understand and quantify.

Many lumped models can simulate with good efficiency the runoff dynamics at the outlet of the catchment (Beven, 2001). However, in many tropical regions the runoff time series are not directly comparable with the precipitation in the catchment mainly due to orography (Buytaert et al., 2010), thus evidently limiting the information about internal hydrological processes (Winsemius et al., 2006), particularly for distributed modelling. To overcome this challenge, a more complex distributed model approach is needed to realistically describe the first-order controls and channel processes in the catchment (Scherrer & Naef, 2003; Uhlenbrook, 2004). The first-order controls are the hydrological processes that describe lateral water fluxes based on the soil properties, land use and land cover (Uhlenbrook et al., 2004).

Within the diverse list of available conceptual hydrological models, three models deal with the complexity of the soil routine and the physical properties to estimate the runoff generation processes. For example, WaSiM ETH (Water balance

Simulation Model) is a spatially distributed, process and grid-based hydrological catchment model developed to simulate the water balance in mountainous regions (Schulla & Jasper, 2007). Version 2 of this model uses the Richards-equation for unsaturated soil zone and the Laplacian advection-dispersion equation for the saturated zones. The HBV-96 model is the distributed version of the conceptual model HBV (Lindström et al., 1997). The model uses a sub-basin division with specific elevation bands, vegetation coverage and snow classes. The HBV-96 represents the land-phase of the hydrological cycle by three main components: snow routine, soil routine and runoff response; unfortunately it does not contain a routing scheme for stream runoff. Finally, the TAC^D (Tracer Aided Catchment model, distributed) (Roser, 2001) model is somewhat comparable with the two above mentioned models and consist of several sequentially linked routines. The storages in neighbouring cells are connected through lateral fluxes and different storage types within the same cell are connected by vertical fluxes. The type of storage in a cell and the specific parameters are defined by the hydrological unit type. In addition, the TAC^D uses the kinematic wave routing to estimate the overland flow. The latter, generates runoff in each cell and adds to the kinematic wave reservoir at the end of the time step. The novelty with TAC^D lays in the runoff generation routine described in several hydrological unit types, in which the surface water flux is routed through two or three storages interconnected via vertical or lateral fluxes and therefore performs a more detailed description of the runoff generation processes in each cell.

The representation of the hydrological complexity of Los Crespos-Humbolt catchment is still a challenge. Studies in the same catchment using biophysical models have failed to estimate the water balance due to the limitation in the calculation of the soil moisture component (Minaya et al., 2016a). In this regard, we would like to enhance our understanding of the spatial runoff generation processes in the catchment through the application of the TAC^D model. The existing catchment model TAC^D (Tracer Aided Catchment model, distributed) has been applied successfully in several studies (Uhlenbrook, 1999; Roser, 2001; Ott, 2002; Johst, 2003; Sieber, 2003; Tilch et al., 2003; aus der Beek, 2004) with a high degree of physical agreement.

The aim of this chapter is to understand the hydrological processes and the homogeneity of the catchment through an integrated approach that takes into account the catchment biophysical features and high spatial distribution of the meteorological data. These will be done in three steps: 1) Identification of the hydrological units that hold the same runoff generation process within the catchment, 2) Implementation of the TAC^D model in a distributed 30 x 30 m2 on a daily basis, and 3) Calibration of the model through the use of a simple genetic algorithm to optimize a large number of parameters that describe the runoff processes of the different hydrological units identified in the TAC^D.

7.2 Methods and data

The study area is the same described earlier in Chapter 2 (sections 2.2 to 2.4).

7.2.1 TAC model

The TAC^D (Tracer Aided Catchment model, Distributed) reproduces the water fluxes in a distributed, process-oriented way (Uhlenbrook, 1999) using the PCRaster environmental modeling language (Karssenberg et al., 2010; Wesseling et al., 1996) to link the dynamic, space related operations and time series of data for distributed hydrological modeling. We used the last version of the TAC^D model developed by Wissmeier (2005) who successfully corrected the code for several bugs and an appropriate hydrological formulation. This last version uses a solute transport that is tied with the water fluxes; however for the current study we will only use the formulation without the solute transport component. Lateral fluxes use the simple differential equation of a linear storage unit as the governing equation (Eq. (7-1)); whilst vertical fluxes are represented by constant amounts of percolating water, which are independent from storage levels until storages are empty (Eq. (7-2)).

$$-\frac{dV}{dt} = k^* x\, V = Q^* \tag{7-1}$$

where k^* is the storage coefficient (1/time step), V is the storage level (mm) and Q^* is the flux (mm/time step).

$$-\frac{dV}{dt} = c \tag{7-2}$$

where c is the constant parameter for vertical flux (mm/time step)

The flow can be reported in stream discharges at the outlet as well as at different locations specified within the catchment. The innovation of this model is the runoff generation routine developed notably for mountainous areas, where the flow direction is entirely regulated by the hill slopes. The TAC^D uses different storage types in the same cell for the runoff generation; vertically cells are divided in different layers that represent single storages and these are interconnected via vertical (to different storages in the same cell) or lateral (to neighboring cells) fluxes

(Uhlenbrook et al., 2004; Uhlenbrook and Sieber, 2005). The model routes the water using the single-flow direction algorithm D8 (O' Callaghan and Mark, 1984).

The modular structure of the TAC^D is shown in Figure 7-1, which also includes the models that were used to prepare some of the input data and the model that was used to calibrate the simulations. The recent modifications of the model have added several advantages in its applicability compared to other similar models. For instance, the implementation of a control tool to assess the model for any violations of mass conservation, modification of the slope factor that limits the outflows to the actual storage levels, actual evaporation from the soil routine has been limited to its water content, among others (Wissmeier, 2005). Calibration analyses were performed using a genetic algorithm (GA) calibration approach for optimization (Scrucca, 2012) with one objective function which is the minimum RMSE (Figure 7-1).

The TAC^D starts with an initialization run to reach stability to determine the initial conditions of soil moisture and storage levels of saturated and unsaturated areas. The complexity of the model lays in the identification of the hydrological functional units that at a certain extend require an empirical approach by overlaying geographical, physical and geological features in addition to personal user knowledge to spatially delineate areas with the same dominating processes (Uhlenbrook et al., 2004; Wissmeier, 2005). Thus, the predominant generation processes are subject to large uncertainties due to the high heterogeneity throughout this grassland dominated catchment.

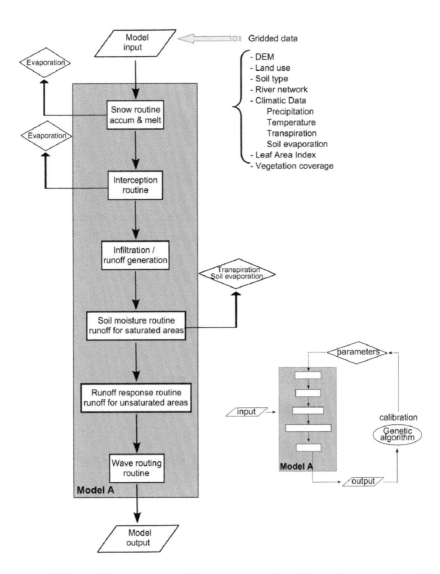

Figure 7-1 a) TAC^D model structure, round shapes are external models used to prepare input data, b) schematic diagram of the calibration using a genetic algorithm.

7.2.1.1 Model inputs

Precipitation

Rain gauge stations located inside the basin measured precipitation at different frequency (Figure 2-6). We selected four stations that had a robust data set and proceeded to fill the small gaps (from days to maximum 2 weeks) with a model tree approach that included a training and testing procedure. The filled data was cross-checked and corrected with the monthly data of the closest monthly rain gauge. A combined inverse distance method and an elevation gradient to regionalize to the catchment scale was applied by using a tool for grid-based interpolation of hydrological variables Hykit (Maskey, 2013). The equation used for the interpolation was:

$$\hat{P}_k = W_D \sum_{i=1}^{N} \frac{1}{D} w(d)_i p_i + W_Z \sum_{i=1}^{N} \frac{1}{Z} w(z)_i p_i \qquad (7\text{-}3)$$

where $\hat{P}_k$ (mm/day) is the interpolated precipitation in a grid cell k, W_D and W_Z are the factors for distance and elevation, respectively, p_i is the precipitation (mm/day) of the ith rain gauge station, N is the number of rain gauge, $w(d)_i$ and $w(z)_i$ are the individual gauge weighting factors for distance and elevation (Eq. 7-4 and Eq. 7-5), respectively, D and Z are the normalization quantities given by the sum of individual weighting factors for all the rain gauges used in the interpolation.

$$w(d) = \frac{1}{d^a} \qquad \qquad \text{for } d>0 \qquad (7\text{-}4)$$

$$w(z) = \begin{cases} \frac{1}{z_{min}^b} & \text{for } z \leq z_{min} \\ \frac{1}{z^b} & \text{for } z_{min} < z < z_{max} \\ 0 & \text{for } z \geq z_{max} \end{cases} \qquad (7\text{-}5)$$

where d is the distance in km between the cell grid k and the rain gauge station used for interpolation, z is the absolute elevation difference in m between the cell grid and the rain gauge station, a and b are the exponents factors for distance and elevation, respectively, z_{min} and z_{max} (m) are the minimum and maximum limiting values of elevation differences (Daly et al., 2002).

We tested several combinations and selected the one with the smallest root mean square error (RMSE).

Temperature

Temperature (°C) was obtained from the meteorological stations (Figure 2-6) and correlated with elevation in order to obtain a gradient (the difference in temperature and elevation). Then the gradient was applied to fill gaps and to interpolate the temperature along an altitudinal gradient for all grid cells in the catchment.

Transpiration and soil evaporation

The transpiration and the soil evaporation was calculated with the Penman-Monteith equation (Monteith, 1973) using stomatal conductance (Running and Coughlan, 1988). This equation uses features of a specific surface and current meteorological data (incoming radiation, vapor pressure deficit, air temperature and air pressure) to estimate an instantaneous heat balance of an object. The gridded data for transpiration and soil evaporation were taken from Chapter 4 that used the ecosystem process-based model BIOME-BGC (Minaya et al., 2016).

Leaf Area Index

Distributed monthly leaf area index (LAI) for a year (January to December) was estimated from daily LAI from a 12-year period. First, daily LAI was averaged to monthly and then averaged for a specific month, for instance all monthly LAI values from January were averaged to have one representative value of all Januaries.

Soil properties and land cover

In Chapter 3, it was demonstrated that the texture for vegetated soils was related to an altitudinal gradient, where sandy soils were dominant at higher elevations (> 4500 m a.s.l.) whereas silty soils were mostly located at lower elevations (Minaya et al., 2015a).

The slopes in the low and mid catchment are moderate with values up to 15° and increase up to 30° towards the moraine at higher elevations. In Chapter 2, it was stated that these types of vegetation in the *páramo* ecosystem are able to catch low energy rain, drizzle and fog moisture on their leaves, which can significantly conduct rainwater directly to the soils (Crockford and Richardson, 2000; Foot and Morgan, 2005; Janeau et al., 2015).

Hydrological data

Water levels were measured at the outlet of Los Crespos and Humbolt subcatchments. Unfortunately, the data is not continuous showing a lot of missing gaps that were difficult to fill with simple methods. We took only the reliable data that could be calculated with the rating curve (from 2005 to 2008). The storage volumes were obtained by the initialization of the model (915-day simulation) to reach realistic conditions.

7.2.1.2 Identification of Hydrological Units

Runoff generation and water routing

We used empirical approach to delineate areas with the same dominating processes taken into account the characteristics of land use, soil properties, topography, geology and drainage network as carried out in previous experimental investigations (Tilch et al., 2002; Uhlenbrook and Leibundgut, 2002). Besides the above mentioned criteria, for some hydrological units we used data of environmental tracers from Chapter 6 to infer information about interflow and resurgence of water downstream. Additionally, saturated areas and drift cover were mapped during the same field survey for the study site. With all the information, we delineated hydrological units for which runoff behavior can be comparable and assumed to be identical. The spatialization of the runoff generation types in the catchment was done using the PCRaster environmental modelling language (Karssenberg et al., 2010).

A schematic sketch of a hill slope and the conceptualization of the runoff generation process for each hydrological unit type that are connected within the TAC^D model is shown in Figure 7-2. These reservoirs systems are connected sequentially by overflowing processes based on the reservoir content S (mm), the storage coefficient k (h-1) and the local slope at each raster cell.

The lateral outflows that connect the upper storages (s_US and s_MTD) are named sQ_US for the HU type 2 to HU type 5 and sQ_SOF for HU type 6. Likewise, the lateral outflow that connect the lower storages is sQ_LS and sQ_GW for the groundwater storage. The upper storage connects to the lower by the vertical outflow sStorageLeak and viceversa by the inflow from the full lower storage sQ_LSfull. The lower storage connects to the groundwater storage the vertical outflow sToGroundwater.

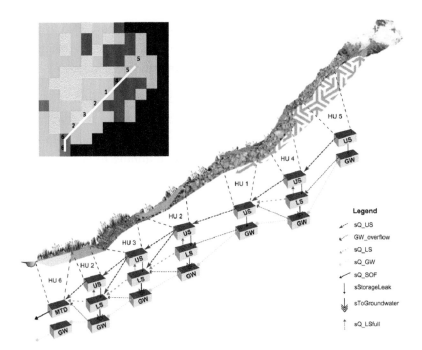

Figure 7-2 a) Small section of the runoff generation map, routing from HU type 5 to HU type 6 and passing by HU type 1 to 4; b) Schematic conceptualization of the runoff processes and their interconnection via lateral fluxes with other neighbour HU type cells.

Water routing

The runoff routing of the stream network used the kinematic wave approach (Chow et al., 1988), which is based on the combination of momentum and mass conservation laws, predefined as a PCRaster function kinematic. The simulation time step was 8640 s for a cell size of 30 x 30 m² and it depends on space and time discretization. The routing module used the digital elevation model (DEM) to calculate slope. The channel length of the stream network was set to 32.20 m, taking also the slope of those cells. The width of the stream network was recorded during a field survey and it varies from 0.4 to 2m. The Manning's coefficient for the streams was set to 0.05 for small mountainous streams with a bottom of cobble and gravel (Chow, 1959).

7.2.1.3 Model parameterization and calibration

A TAC^D aggregated model that uses a single parameter set for all hydrological units was used as a first attempt to estimate some parameters and check their sensitivity. For this, we started with rough estimations based on other similar basins and from literature review (Table 7-2). Maximum and minimum values were derived from similar studies in mountainous basins (Sieber and Uhlenbrook, 2005; Uhlenbrook et al., 2004) and from the HBV manual calibration to obtain the best agreement between the simulated and observed flow at the outlet of the study catchment.

Calibration analyses were performed using a genetic algorithm (GA) calibration approach for optimization (Scrucca, 2012) with one objective function which is the minimum RMSE. The GA optimized the search based on a natural selection processes similar to the one of biological evolution aiming to evolve in an optimal solution. The GA is meant to solve the optimization problem that minimizes the objective function as shown in Eq. 7.6 and Eq. 7.7.

$$RMSE = \frac{1}{n}\sqrt{\sum_{i=1}^{n}(\bar{x} - x)^2} \tag{7-6}$$

$$\min_{x} RMSE \approx 0 \tag{7-7}$$

where n is the number of days, x is the observed discharge value (m³/s), $\bar{x}$ is the simulated value (m³/s). This latter is a function of the 33 parameters that will be calibrated (Eq. 7.8).

$$\bar{x} = f(P) \tag{7-8}$$

where $P = P_1, P_2, \dots, P_{33}$, these parameters are indicated in the Appendix section.

This optimization problem was setup to run in parallel environment due to the high computational load of the distributed model. This was implemented with a connection from Matlab to the PCR-Calc module in DOS and ran in an 8-core server. The ideal procedure for GA is to have enough diversity in a population; our problem is composed by 33 parameters with a relative clear range of possible values per parameter. Therefore the GA was setup to cover more the input space, and to allow less generations of the population. In general terms, we chose a size of 1000 random numbers within the range provided and with 100 generations. Finally a comparison between the observed and simulated data obtained from the best fit model was done to validate the accuracy and applicability of the model.

7.2.1.4 Model comparison for soil water content

A comparison of the soil moisture estimated with the biogeochemical model BIOME-BGC (refer to Chapter 4) and the hydrological model TAC^D was performed to identify possible patterns and add understanding to the hydrological processes in the soil component.

7.3 Results and discussion

7.3.1 Meteorological information

The best weighted average to calculate the precipitation in each cell at a daily time step was obtained with 70% from the inverse distance weighting and 30% from the elevation. Location of the rain cell is more important than elevation; however, the spatial distribution was slightly compromised due to the lack of a long time series data of wind velocity that could have helped to improve the results in the mountainous regions.

Temperature (°C) was highly correlated with elevation. Two clear gradients were found, -0.42 °C for every 100 m for elevations between 4000 to 4785 m a.s.l. and -1.10 °C for every 100 m for elevations greater than 4785 m a.s.l., basically the ones located on the glacier.

We obtained daily distributed maps for all hydro-meteorological input data and monthly maps for Leaf Area Index.

7.3.2 Hydrological units

Each cell in the catchment belongs to one of the six hydrological units identified, and they mainly represent the conceptual composition of the storage levels, coefficients and fluxes (Wissmeier, 2005). The hydrological units (HU) shown in Figure 7-3 are described as follows:

HU (1): areas with stratified soil (around 50% sand and 50% clay) with slopes up to 12°. It has only two storages, i.e. upper (sUS) and groundwater (sGW).

HU (2): areas with stratified soil and a base layer of stable stones and sandy loam. Slopes between 1.5° to 12° with delayed interflow. Lateral flow is dominant in the main layer and it consists of three storages, i.e. the upper storage (sUs), the lower storage (sLS), and the groundwater storage (sGW).

HU (3): areas with sandy soils with coarse boulders and loose stones. Slopes between 12° to 20° with delayed interflow. It has three storages similarly to HU (2) with high lateral hydraulic conductivities.

HU (4): areas with block and boulder layers with the presence of sandy soils. Located in very steep areas > 20°, geological matrix identified as Hialina lava and Pleistocene lavas (Hall et al., 2012), represented in three storages (same as HU (2)) with quick lateral interflow.

HU (5): areas located in the moraines and on hillslopes with a mixed distribution of soil texture. These units are located at the toe of the glacier and might have a significant storage volume that contributes to base flow. The water percolates to deeper zones to the fractured bedrock contributing slowly further downstream by resurgence (Cauvy-Fraunié et al., 2013). It is represented by two storages (same as HU (1)) with extreme delayed stream flow.

HU (6): areas with saturated overland flow as predominant runoff generation process. These areas are constantly wet, located in riparian zones, near springs with a very low slope < 1°. During the field survey, saturation was observed in vegetated areas dominated by cushions. This type is represented by two storages. The first storage ($sMTD$) has a low maximum storage capacity (around 30 mm) and only little water can be stored before overflow is triggered. The second is a ground water storage (sGW) that has no connection with the earlier one and it is supplied only by lateral inflow from its neighboring cell. The stream network cells are also part of this HU.

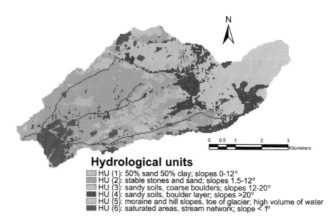

Hydrological units
HU (1): 50% sand 50% clay; slopes 0-12°
HU (2): stable stones and sand; slopes 1.5-12°
HU (3): sandy soils, coarse boulders; slopes 12-20°
HU (4): sandy soils, boulder layer; slopes >20°
HU (5): moraine and hill slopes, toe of glacier; high volume of water
HU (6): saturated areas, stream network; slope < 1°

Figure 7-3 Hydrological unit types that show dominant runoff generation processes in the Los Crespos- Humboldt basin, Ecuador.

7.3.3 Model parameterization

The aggregated TAC^D model for the period 20.11.2006 - 10.09.2008 was used to provide good estimations of the parameters that later were used for the TAC^D model. Figure 7-4 shows the results obtained with the aggregated TAC^D model to determine the first rough estimation of parameters that were not further optimized. In general, the flow dynamics reproduced fair enough with some underestimations throughout the flow period. As explained earlier the intention of this approach was not to have a perfect agreement between observed and simulated flows but rather to find an approximation of the parameters that will be later optimized for each hydrological unit. The RMSE calculated in this first approach is 0.061.

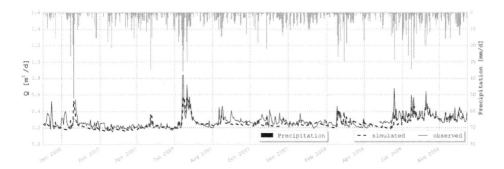

Figure 7-4 Results of the discharge simulation using the aggregated TAC^D Model for the initial estimation of parameters.

7.3.4 Model simulation

The TAC^D model ran first an initialization period of 915 time steps in order to find equilibrium conditions and estimate realistic values of the initial water storage conditions and states; and it was not taken into consideration for further analysis during calibration. The initialization run gave new conditions of upper, lower and groundwater storage as well as soil moisture for all the hydrological units. Initial conditions are crucial for the overall simulations and therefore they should take into consideration the spatial distribution of precipitation and evaporation patterns. The model simulations tested in other studies (Wissmeier and Uhlenbrook, 2007) have demonstrated to be highly sensitive to the initial conditions. It is worth to mention that an appropriate quantification of water retention in the soils is crucial for the assessment of the vegetation of these Andean highlands ecosystems.

A period of 1371 time steps was selected for the simulation that corresponds to the period between 31-03-2005 and 31-12-2008. The lack of a strong seasonality in this high-altitude catchment allows the period to be representative. During the optimization, we selected the parameters that gave the best performance for the entire period and not specifically fitting peaks or extreme events.

Figure 7-5 shows the correlation between observed and simulated discharge for the TAC^D aggregated model used during the parameterization process and the TAC^D model used for the simulation period. During the parameterization, the model gave us results with a reasonable agreement with the observed flows with a correlation of 0.49, RMSE of 0.06 and a positive NASH efficient coefficient of 0.40 (Table 7-1). Later during calibration of the 33 parameters with the genetic algorithm, the simulation results lowered the level of agreement that we had before (Figure 7-5b and Table 7-1) and therefore it does not provide adequate representation of the hydrological system. By comparing these results, it is noticeable the complexity of the model during the calibration of 33 parameters that represent the processes of six hydrological units within the catchment.

The TAC^D model tried to represent a physical description of water flow processes in mountainous catchments as evidenced in other catchment studies (Ott and Uhlenbrook, 2004; Sieber and Uhlenbrook, 2005; Wissmeier, 2005). However, the errors calculated from the difference between the simulated and observed runoff might be attributed not only to the runoff generation descriptions but it could be also associated to the uncertainties in the input spatial distribution of soil evaporation, transpiration and precipitation and the lack of a larger time series input data for the model initialization.

These high-altitude catchments are subject to large spatial climatic variability as explicitly discussed in other studies in the same region (Buytaert and Beven, 2011; Buytaert et al., 2006b; Celleri et al., 2007). Despite the efforts done in this study to estimate the spatial distribution of precipitation with necessary confidence, it did not consider the wind as a co-variant due to the lack of a robust set of data and uncertainties associated with its regionalization. This drawback might have influenced the precipitation patterns horizontally as well as along an altitudinal gradient for each cell and therefore the volume of runoff at the outlet of the catchment.

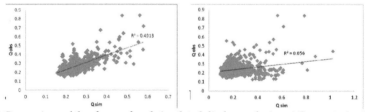

Figure 7-5 Comparison of the observed and simulated discharge during a) Parameterization using a TAC^D aggregated, b) Calibration using TAC^D.

Table 7-1 Evaluation of the aggregated TAC^D used and the TAC^D models.

	Coefficient correlation	RMSE	Coefficient Efficiency
TAC^D aggregated	0.49	0.06	0.40
TAC^D	0.06	0.12	-1.11

In the following section, we will perform an uncertainty analysis of the parameters to check if the model is reliable or not.

7.3.5 Uncertainty obtained from the optimization process

7.3.5.1 Parameter analysis

To evaluate the uncertainty of the optimization results we explored the parameter space via a graphical representation of the errors. For this, a scatter plot shows how the RMSE changes with change in each of the parameters (Figure 7-6). The points in the graph can be interpreted by the way the optimization algorithm was performed, while the GA increases accuracy the points will narrow down into sub-regions. Regions with higher concentration of points would have reached a more stable value of the parameter than others.

Figure 7-6 shows the 33 parameters that were changing at the same time in every run. The results show that parameters related to the temperature threshold for snow fall and temperature for snow melt within the snow routine reached stability, showing that these ranges did not narrow more due to the early stopping criteria defined in the GA. In the soil routine, the field capacity parameters for all hydrological units show sensitivity and therefore they might represent a high uncertainty in the model. Most of the parameters in the runoff generation routine seem to be very homogeneous with high dispersion of values in the performance of the model except the parameter MTD, which represents the maximum height of micro-topographic depression. This parameter defines the first storage in the hydrological units with saturated overland flow, areas constantly wet and therefore a very small amount of water can be store before it overflows. The main issue with this type of conceptualized models is the equifinality, which is a problem of the model and not necessarily with the calibration. The equifinality theory suggests that several suitable parameter sets exist for achieving our objective function, which is to minimize the error measured through the root mean square error (Beven, 2006).

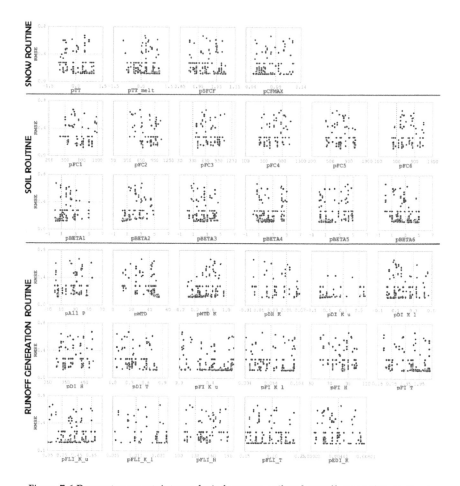

Figure 7-6 Parameter uncertainty analysis for snow, soil and runoff generation routines.

7.3.5.2 Confidence bounds

The confidence bounds of the model parameters during calibration are shown in Figure 7-7 (a). A simple approach was carried out to be able to generate the confidence bounds of the optimized model using the data generated during optimization. Therefore, these confidence intervals are limited to the most likely parameters estimated by the GA. The GA generates the "most likely" solutions aiming to minimize the error in the objective function. Figure 7-7 (b) shows a blue line that represents the optimum solution that gave the lowest RMSE, it can be seen that is skewed to the upper boundary of the solution, which means that it has a large uncertainty in the estimation of low discharges.

There are several sources of uncertainty such as model structure, input data and model parameter uncertainties among others. They all can have the same probability to introduce error in the prediction of a variable. However, it is difficult to differentiate the sources of error that contribute to the uncertainty of the results and they are outside of the scope of the present study.

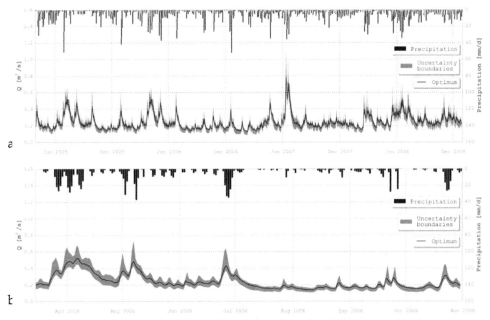

Figure 7-7 a) Time series of the discharge with confidence bounds of 95% for the calibration, (b) zoom for the day 1000 to 1300.

Although it seems that there is a decent representation of the confidence bounds, the uncertainty analysis from the optimization processes revealed that the parameters that represent the hydrological processes for each hydrological unit have an equifinality issue that could not fully explain the interval processes such as soil moisture. However, in the following section as an example we compare the soil moisture estimated with the biogeochemical model BIOME-BGC (calculated in Chapter 4) versus the estimations from the TAC^D.

7.3.6 Comparison of water content in soil

Usually the spatial and temporal variability of soil moisture is due to heterogeneity in soil properties, land cover, topography, irregular distribution of precipitation, elevation gradient, temperature and evapotranspiration that considered not only the radiative forces but also the vegetation properties. However, different models have different perspectives for the representation and estimation of this component. The soil moisture of the BIOME-BGC is conditioned to both the effective soil depth, where the roots can absorb the water, and the soil texture information (% sand, silt and clay) (Cosby et al., 1984). Whilst, the TAC^D conceptualizes the soil routine by describing the infiltration and percolation through the soil layer using an empirical exponential equation (Bergstrom, 1995). In the TAC^D, the parameters to determine the soil moisture such as field capacity and beta are determined during calibration for each of the hydrological units.

Figure 7-8 shows the response of soil moisture estimated with the biogeochemical model BIOME-BGC and the hydrological model TAC^D to the same precipitation and evapotranspiration. The soil moisture estimated with the BIOME-BGC responds rapidly to the peaks of precipitation whiles the other one apparently dampers the content of water in the soil. However, these results cannot be taken for further analysis since the capability of the TAC^D model to predict adequately the discharge was unsuccessful.

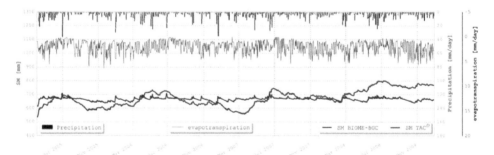

Figure 7-8 Comparison of soil moisture estimations by the biogeochemical model BIOME-BGC and the hydrological model TAC^D.

7.4 Conclusions

Although the TAC^D has the capability to integrate hydrological heterogeneities such as evapotranspiration of the vegetation, hillslope response, overland and interflow water fluxes and runoff response of saturated areas and disconnected wetlands, it failed to simulate the runoff at the outlet of the catchment. It is worth mentioning, that the intention was far from having a reliable prediction of runoff, but to have a better understanding of the physical hydrological processes. Therefore, the results of the TAC^D together with the parameters of the soil routine cannot provide adequate information of the soil hydrological processes represented in the system.

We performed a calibration using a genetic algorithm to obtain a comprehensive and detailed model with grid cells of 30 x 30 m that could help understand the model structure and effectiveness of the chosen parameters, especially the ones related to water fluxes and storage. However, the results were homogeneous and there was not a clear difference in the sensitivity of the parameters except for the parameter MTD, which defines the storage of areas that are constantly wet with a very low slope. Therefore, as a further step would be to analyze another dimension and check the relationship that one parameter has with another one.

The model implementation has generated concerns about the conceptualization and description of the runoff generation processes and therefore the physical processes of the water fluxes and states represented in the system. An adequate representation will help to understand and to assess the distribution of available soil moisture in the catchment which is critical to soil chemical processes, especially for nitrogen fixation that contributes to the availability of nutrients in the soil. Soil moisture plays an important role in the potential regeneration of *páramo* vegetation which is characterized by soils with low availability of nitrogen and phosphorus for plant uptake as a result of their volcanic origin.

Undoubtedly, a better description of the runoff generation processes can lead to a more reliable representation of the hydrological processes, which in turn reduce the model uncertainties. However, in order to have a complete assessment of the usefulness of the process-oriented catchment model TAC^D in mountainous regions in the Andean *páramos*, an uncertainty analysis of the model processes as well as the input data should be further analyzed. Future research should include long-term time series of tracers to fully test the solute transport module of the TAC^D model and reduce data uncertainty while improving reliability of modelling results. In addition to that, glacier dynamics should also be integrated since it plays an important role in terms of soil moisture estimation.

A correct representation of the hydrological processes in the *páramos* is needed to quantify the ecosystem services in terms of water regulation and carbon sequestration. This is necessary for sustainable management of these mountainous regions in the Andean *páramos*.

Appendix 7-A

Table 7-2 Parameters used for the TAC^D model, minimum and maximum ranges used for calibration and the optimal parameters for this case study.

Parameter	Unit	Explanation	Min.	Max.	Estimation method	Case study
Snow routine						
pTT	°C	Temperature threshold for snow fall	-1	1	calibration	-0.78
pTT_melt	°C	Temperature for snow melt	-1	1	calibration	-0.54
pSFCF	-	Snow fall correction factor	0.9	1.1	calibration	1.08
pCFMAX	mm/(°C*hr)	Degree-hour factor	0.04	0.125	calibration	0.15
pCWH	-	Water holding capacity	0.1		(Sieber and Uhlenbrook, 2005)	0.1
pCFR	-	Refreezing factor	0.05		(Sieber and Uhlenbrook, 2005)	0.05
Soil routine						
pLP	-	Reduction of potential evapotranspiration	0.6		(Sieber and Uhlenbrook, 2005)	0.6
pFC1	mm	Field capacity at HU type 1	500	1000	calibration	979.58
pFC2	mm	Field capacity at HU type 2	400	1100	calibration	1027.70
pFC3	mm	Field capacity at HU type 3	500	900	calibration	813.0
pFC4	mm	Field capacity at HU type 4	300	500	calibration	490.93
pFC5	mm	Field capacity at HU type 5	500	1100	calibration	1097.30
pFC6	mm	Field capacity at HU type 6	300	500	calibration	495.28
pBETA1	-	Beta parameter at HU type 1	1	5	calibration	4.93
pBETA2	-	Beta parameter at HU type 2	0.75	5	calibration	4.97
pBETA3	-	Beta parameter at HU type 3	0.5	5	calibration	4.93
pBETA4	-	Beta parameter at HU type 4	0.75	5	calibration	4.30
pBETA5	-	Beta parameter at HU type 5	1	5	calibration	4.97
pBETA6	-	Beta parameter at HU type 6	1	5	calibration	0.63
Runoff generation routine						
pAll_P	mm/hr	Percolation to deeper ground water	0.04	57.6	calibration	41.49
pMTD	mm	Maximum height of "micro-topographic depression"(MTD)	10	50	calibration	42.15
pMTD_K	h^{-1}	Storage coefficient for HU type 6	0.0025	0.2	calibration	0.06
pDH_K	h^{-1}	Storage coefficient (h^{-1}) for HU type 1 (deep percolation in high areas)	0.00025	0.08	calibration	0.07
pDI_K_u	h^{-1}	Upper zone storage coefficient for HU type 2 (delayed Interflow)	0.006	0.1	calibration	0.01
pDI_K_l	h^{-1}	Lower zone storage coefficient for HU type 2	0.00125	0.02	calibration	0.10
pDI_H	mm	Maximum storage capacity of lower zone at HU type 2	300	500	calibration	418.15

pDI_T	mm/hr	Percolation to lower storage at HU type 2	0.05	0.8	calibration	0.28
pFI_K_u	h⁻¹	Upper zone storage coefficient for HU type 3 (delayed interflow)	0.005	0.7	calibration	0.05
pFI_K_l	h⁻¹	Lower zone storage coefficient for HU type 3	0.006	0.1	calibration	0.03
pFI_H	mm	Maximum storage capacity of lower zone at HU type 3	60	100	calibration	78.94
pFI_T	mm/hr	Percolation to lower storage at HU type 3	0.15	2.4	calibration	0.79
pFLI_K_u	h⁻¹	Upper zone storage coefficient for HU type 4 (fast, lateral interflow, piston flow)	0.05	0.7	calibration	0.20
pFLI_K_l	h⁻¹	Lower zone storage coefficient for HU type 4	0.002	0.025	calibration	0.04
pFLI_H	mm	Maximum storage capacity of lower zone for HU type 4	110	190	calibration	126.36
pFLI_T	mm/hr	Percolation from upper to lower reservoir at HU type 4	0.15	2.4	calibration	0.20
pEDI_K	h⁻¹	Upper zone storage coefficient for HU type 5, extreme delayed stream flow	0.0005	0.008	calibration	0.0038
pGW_K	h⁻¹	Storage coefficient for hard rock	0.005		(Wissmeier, 2005)	0.005
pUS_H	mm	Maximum storage capacity for all HU	800		(Wissmeier, 2005)	800
pGW_H	mm	Maximum storage capacity for ground water	1000		(Wissmeier, 2005)	1000
pThres	mm	Threshold limit of upper storage for infiltration and exfiltration	500		(Wissmeier, 2005)	500
pBeta	-	Parameter for kinematic wave routing	0.6		(Chow et al., 1988)	0.6
pTimeStep	s	Time step for kinematic wave	8640		Calculated (based on space and time discretization)	8640
pQIni	m³/s	Initial stream flow	0.05		(Wissmeier, 2005)	0.05
pWaterDepthIni	m	Initial water depth	0.05		(Wissmeier, 2005)	0.05

*300 trout are needed to support 1 man for a year. The trout in
turn, must consume 90,000 frogs that must consume 27 million
grasshoppers that live off 1,000 tons of grass (G. Tyler Miller)*

8

ECOSYSTEM SERVICES ASSESSMENT IN A PÁRAMO SYSTEM

Many efforts have been made to identify the human benefits that can be obtained from ecosystem goods and services. However, to the best of our knowledge, none of them have considered that landscapes differ in their capabilities to provide these services. In this chapter, we identify the main ecosystem services provided by a *páramo* system using the Ecosystem Service Cascade approach that links ecological structures and processes to the benefits that people can get from such ecosystem. To accomplish this, we use the information of all previous results of carbon stocks and water resources availability in the catchment study, which include detailed landscape information to represent the heterogeneous distribution of the ecosystem properties such as plant functional types, site/soil parameters and daily meteorology. The aim is to build indicators needed for evidence-based policy making that can be used by public institutions that are legally responsible for the management of natural resources. The outcome of this chapter can be used as complementary information for an adequate assessment and fair evaluation of conservation initiatives such as payment for ecosystem services that is now being applied in the region.

This chapter is based on:

Minaya, V., Gonzalez-Angarita, A., Corzo, G., van der Kwast, J., and Mynett, A. E.: Assessment of Regulation and Maintenance Ecosystem Services in an Ecuadorian *páramos*, in preparation, 2016.

8.1 Introduction

Ecosystem services have been widely recognized as significant due to its importance for human welfare (d'Arge et al., 1997). These services own an intrinsic and utilitarian value (Liu et al., 2010) which benefits directly or indirectly all stakeholders (Hein et al., 2006), e.g. local community, local government, national and international development organizations, among others. Costanza et al. (2014) established that the contribution of the ecosystem services can be twice as much to human well-being as the global GDP (gross domestic product). Several indicators and metrics have been developed to qualify and quantify the different classifications of ecosystem services (Walpole et al., 2011). However, there was an urgent need to unify the criteria, compare experiences and establish basic concepts to define a standardized ecosystem service classification (Haines-Yong and Potschin, 2013). The Common International Classification of Ecosystem Services (CICES) collected the different perspectives and presented a classification structure based on a 3-digit level, namely: (1) provisioning, (2) regulating and maintenance and (3) cultural services. Its main characteristics are extensively described in the study of Haines-Young & Potschin (2013).

Neotropical high altitudinal ecosystems in the Andean Region are known as *páramos* and they play a significant role in rainfall interception, water storage and regulation processes (Armijos and De Bièvre, 2012; Mena et al., 2000) as well as water supply for over 10 million people in the northern Andes (Buytaert et al., 2006a). In Ecuador, the *páramos* are located at elevations between 3000 and 4700 m a.s.l. and cover over 13.370 km² (Cuesta et al., 2012; Hofstede et al., 2003). Its altitudinal location allows their interaction with the tectonic uplift as well as volcanic and glacial activity, which are the main driving factors shaping the current landscape (Hribljan et al., 2016). The soils of these ecosystems are characterized by a high water retention capacity and abundant organic matter (Hofstede, 1997; Luteyn and Balslev, 1992; Minaya et al., 2015a; Podwojewski, 1999), which enhance the ecosystem services that they provide. The identified services provided by the Ecuadorian *páramos* are water supply, carbon storage and biodiversity. All of them are essential for livelihoods of local communities (Buytaert et al., 2011; Medina and Mena-Vásconez, 2001; Myers et al., 2000). However, these high-altitude ecosystems are being affected by several factors such as land-use change, agriculture and floriculture expansion, burning of grasslands, opening of highways, mining, and changes associated to the global climate (Armijos and De Bièvre, 2012). These have resulted in a *páramo* loss of 1826.6 Ha during the period 1988 – 2007, which represents 20% of the area registered in 1988 (Wigmore and Gao, 2014). However there is not a clear understanding regarding the long term effects of these activities due to their unsustainable development (Benavides, 2014; Buytaert et al., 2006a; Hribljan et al., 2016).

Since the term "ecosystem service" emerged in 1966 (King, 1966), around 15500 articles and review publications have been identified (Scopus Database until December 2015); these studies highlight the importance to bridge the gap between the environment and the human well-being as well as supporting conservation and sustainable use (Potschin and Haines-Yong, 2011). However, only about 28 publications are related to this topic in the Andean grasslands or *páramo* ecosystems (Figure 8-1).

In Ecuador, there are a few studies related to ecosystem services quantification like carbon storage and the implications in the ecosystem services in the *páramos* as well as assessment for payment for ecosystem services (Farley et al., 2011; Farley et al., 2013; Farley et al., 2004; Hribljan et al., 2016). Some initiatives were created to encourage the conservation of these ecosystems while living conditions for the nearby community are being improved. SocioPáramo, is a nationwide coverage program that protects 800000 Ha of *páramo* through the compensation for ecosystem services such as carbon sequestration, water regulation and biodiversity and at the same time improving the quality of life of people currently living in poverty (Farley et al., 2011; Farley et al., 2013). The program developed a payment structure strategy appealing to small landowners by paying a certain amount per hectare (~30 USD/Ha) and decreasing the payment proportionally for larger areas. However, the assessment does not take into account a quantification method and the difference of the landscape's capability to provide ecosystem services. In this chapter, we are therefore proposing a series of environmental services indicators in order to quantify the regulation and maintenance services that the *páramo* ecosystem provides along an altitudinal gradient and for the dominant vegetation in the area.

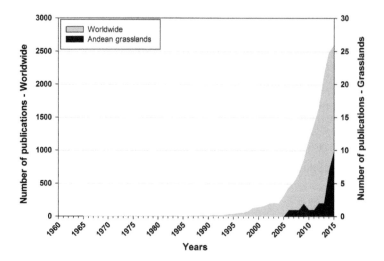

Figure 8-1 Articles and review publications related to the term "ecosystem service" in the Scopus Database (until December 2015)

8.2 Materials and methods

8.2.1 Study area

The case study is the same as described earlier in Chapter 2 (sections 2.2 to 2.4). As demonstrated earlier in Chapter 3, the growth forms had large differences in their carbon, nitrogen concentration and main ecophysiological characteristics along altitudinal gradients (Minaya et al., 2015a). In this regard, the analysis will keep the same analysis at three elevations (R1: 4000-4200 m a.s.l.; R2: 4200-4400 m a.s.l.; R3: 4400-4700 m a.s.l.).

8.2.2 Data availability

The above and below-ground biomass as well as the ecophysiological data were collected from a previous study (Minaya et al., 2015a) carried out in 27 sampling sites representing the three-rank levels of slopes: low (0-7%), moderate (7-27%) and high (>27%) along an altitudinal gradient from 4000 to 4700 m a.s.l. in the same catchment. The meteorological information was collected and interpolated from two weather stations: Humboldt (4010 m a.s.l.) and Los Crespos Morrena (4785 m a.s.l.) located at the outlet and in the upper catchment, respectively (Figure 2-6).

8.2.3 Ecosystem services approach

Haines-Yong & Potschin (2010) developed an Ecosystem cascade diagram to link ecological structures and processes with the benefits that people can get from them including the non-material benefits. Based on this concept, we proposed a similar approach for the *páramo* ecosystem and its biodiversity, which offer regulating and cultural services directly related to the human well-being (Figure 8-2). The identified cultural services that the *páramo* ecosystem provides are based on its physical settings, its unique location and biodiversity, which depend on the in-situ living processes. As a physical setting, the study area includes the southwestern slope of the Antisana icecap, where physical activities such as hiking attract high-mountain travellers.

For the purpose of this study, we will derive quantitative measures to estimate the ecosystem services from the regulating services only with focus on water regulation and carbon sequestration as the most important services within the Regulating and maintenance structure as part of the Common International Classification of Ecosystem Services (CICES) (Haines-Yong and Potschin, 2013). The Regulation & Maintenance section entails all aspects in which living organisms can intervene and regulate the physico-chemical and biological environment including climate at global and local scales that affects the status of human well-being.

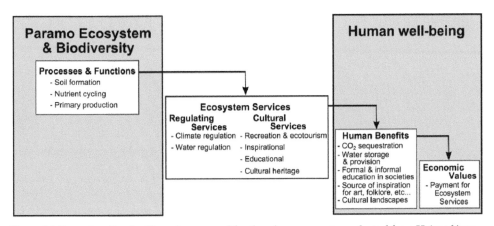

Figure 8-2 Ecosystem Service Cascade proposed for the *páramo* ecosystem adapted from Haines-Yong & Potschin (2010).

To accomplish this we have tested the use of an e-tool to support a more systematic assessment of the ecosystem services that the *páramo* provides. The development of ecosystem service tools have increased in the last 10 years (Bagstad et al., 2013) and several of them have tried to integrate different aspects such as ecology, economy and geography in order to support decision making processes (Daily et al., 2009; Ruhl et al., 2007). During the review of the tools, we have chosen the BIOME-BGC Project Database & Management System build under the frame of the BioVel project (Biodiversity Virtual e-Laboratory Project) from now on it will be called BioVel e-tool. The e-tool uses the biogeochemical and ecophysiological model BIOME-BGC as a model capable of representing ecological and biophysical processes in the ecosystems (White et al., 2000a). The model uses site and soil characteristics, near-surface daily meteorological parameters and ecophysiological features of the *páramos* described in 21 parameters. All the data used for the model simulations were taken from Chapter 3 that analyzed the carbon stocks and biomass and Chapter 4 that estimated the gross primary production along an altitudinal gradient for the three main growth forms of vegetation.

The BioVel e-tool estimates a set of ten ecosystem services indicators (ESIs) based on the terrestrial ecosystems simulations performed by BIOME-BGC (BioVel Portal, 2014). For the *páramo* ecosystem, seven ESIs are applicable to water regulation and carbon sequestration services within the Regulation & Maintenance section (Table 8-1).

Table 8-1 Quantitative Ecosystem Service Indicators for the *páramo* ecosystem (adapted from CICES V4.3).

Section	Division	Group	Class	Human benefit	Ecosystem Service Indicators
Regulation & Maintenance	Mediation of flows	Mass flows	Mass stabilisation and control of erosion rates	Water regulation	*SOILPROT* – sum of living and dead biomass protecting the soil against erosion, kgC/m²/yr
		Liquid flows	Hydrological cycle and water flow maintenance		*DESDWO* – damping of ecosystem daily water outflow, unitless
	Maintenance of physical, chemical, biological conditions	Soil formation and composition	Decomposition and fixing processes	Carbon sequestration	*DECOMP* – litter and coarse woody debris decomposition rate, kgC/m²/yr *HUMIF* – humification in the soil, kgC/m²/yr
		Atmospheric composition and climate regulation	Global climate regulation by reduction of greenhouse gas concentrations		*annNPP* – annual net primary production, kgC/m² *TCS* – total carbon stock, kgC/m²/yr
			Micro and regional climate regulation		*annET* - energy absorption by evapotranspiration, kgH₂O/m²/yr

Table 8-2 shows the daily and annual parameters used in the calculation of each of the ecosystem service indicators.

Table 8-2 Parameters from BIOME-BGC used for the calculation of the Ecosystem Service Indicators.

Ecosystem Service	Class	Indicator	Abb. and units	Parameters	Calculation
Water regulation	Mass stabilization and control of erosion rates	Living and dead biomass protecting the soil against erosion	SOILPROT [kgC/m²/yr]	Carbon in leaves (leafc) Carbon in fine roots (frootc) Carbon in live coarse root (livecrootc) Carbon in dead coarse root (deadcrootc) Carbon in litter (litrc)	SOILPROT(daily) = leafc + frootc + livecrootc + deadcrootc + litrc SOILPROT(yr) = MIN(SOILPROT(daily))
	Water flow maintenance	Damping of Ecosystem Daily Water Outflow	DESDWO [-]	Precipitation (prcp) Soil water lost to runoff and ground water (soilw_outflow)	PRCP_DailyStd(yr) = STD (prcp[yearday1 yearday365]) OUTFLOW_DailyStd(yr) = STD (soilw_outflow[yearday1 … yearday365]) DESDWO(yr) = PRCP_DailyStd(yr) / FLOW_DailyStd(yr)
Carbon sequestration	Decomposition and fixing processes	Litter annual decomposition rate	DECOMP [kgC/m²/yr]	Decomposition of litter to soil (litrc_to_soilc)	DECOMP(yr) = litrc_to_soilc[yr]
		Humification in the soil	HUMIF [kgC/m²/yr]	Decomposition in the soil (soilc_to_soilc)	HUMIF(yr) = soilc_to_soilc[yr]
	Global climate regulation	Annual net primary production	annNPP [kgC/m²/yr]	Daily net primary production (daily_npp)	annNPP(yr) = SUM(daily_npp[yearday1 … yearday365])
		Total average carbon stock	TCS [kgC/m²/yr]	Total of carbon in vegetation, litter and soil (totalc)	TCS(yr) = totalc[yr]
	Regional climate regulation	Annual evapotranspiration	annET [mm/yr]	Canopy evaporation (canopyw_evap) Snow sublimation (snoww_subl) Evaporation of water from soil (soilw_evap) Soil water transpired by canopy (soilw_trans)	annET(yr) = canopyw_evap[yr] + snoww_subl[yr] + soilw_evap[yr] + soilw_trans[yr]

Abbreviations: MIN= minimum; STD= standard deviation; AVG= average; SUM= sum

8.2.4 Data treatment and analysis

The study design involves two independent factors: (i) main growth forms of vegetation (Tussock, Acaulescent Rosette and Cushion), and (ii) three ranked-ordered altitudinal ranges: low catchment (Low, 4000-4200 m a.s.l.), mid catchment (Mid, 4200-4400 m a.s.l.) and high catchment (High, 4400-4600 m a.s.l.). Both factors respond differently in the provision of ecosystem services quantified through ecosystem service indicators. Therefore, the effects of both factors were tested separately for each ESI using one-way analysis of variance (ANOVA) and WELCH variant one-way analysis of means.

The land-cover map based on the pixel classification of the main growth forms of vegetation (30 x 30 m) was used to display the spatial distribution of the ecosystem services both for water regulation and carbon sequestration. For this, the annual results of the ESIs (period 2000 to 2011) were normalized from 0 to 1. Then, the indicators that correspond to the same ecosystem service were added and averaged assuming that each indicator has the same contribution. In this way, each ecosystem service will be mapped with a scale from 0 to 1. An ecosystem service score equal to 1 represents the highest service that can be provided from the *páramo* ecosystem for a specific human benefit. The scores then were extrapolated for the entire catchment.

8.3 Results

8.3.1 Water regulation

8.3.1.1 Mass stabilization and control of erosion rates

The living and dead biomass protecting the soil against erosion assessed by the ESI *SOILPROT* [KgC/m^2/yr] indicator was significantly higher at the higher altitudinal range for all growth forms of vegetation, particularly for tussocks (1.97 kgC/m^2/yr), and decreases towards lower altitudinal ranges (1.40 kgC/m^2/yr) (Figure 8-3a). Focusing only in the higher altitudes, after tussocks, cushions dominated areas offered better protection to erosion than areas covered by acaulescent rosettes (1.21 kgC/m^2/yr vs 1.03 kgC/m^2/yr).

8.3.1.2 Water flow maintenance

The damping of ecosystem daily water outflow assessed by the ESI *DESDWO* [-] indicator did not show any particular trend or significant difference among growth forms of vegetation or altitudinal ranges ($P > 0.05$); however, it showed larger variations in the low altitudinal range in comparison with mid and high altitudinal ranges (Figure 8-3b).

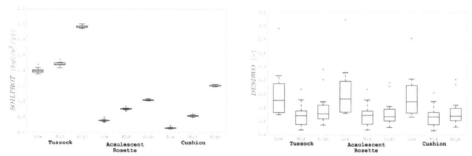

Figure 8-3 Ecosystem Service Indicators of a) Mass stabilization and erosion control rates, and b) Hydrological cycle and water flow maintenance in the *páramos*.

8.3.2 Carbon sequestration

8.3.2.1 Decomposition and fixing processes

The ESI *DECOMP* [KgC/m²/yr] indicator, representing the litter annual decomposition rate, was significantly higher at low altitudinal ranges for tussocks 0.52 [KgC/m²/yr], acaulescent rosettes 0.34 [KgC/m²/yr] and cushions 0.31 [KgC/m²/yr]; and decreased around 50% for tussocks and 25% in cushions and acaulescent rosettes toward higher altitudes (Figure 8-4a). At higher elevation there is no significant difference among the three growth forms of vegetation, the average decomposition rate was around 0.25 [KgC/m²/yr]. A similar pattern was observed with the humification in the soil, estimated as the amount of decomposition in the soil and represented by the ESI *HUMIF* [KgC/m²/yr] indicator. Here, tussocks showed the highest (0.12 KgC/m²/yr) and lowest values (0.05 KgC/m²/yr) at lower and higher altitudes, respectively (Figure 8-4a). For the other growth forms of vegetation, comparing lower vs higher altitudes the average value ranged from 0.08 KgC/m²/yr to 0.06 KgC/m²/yr.

8.3.2.2 Global climate regulation

For global climate regulation by reduction of the greenhouse gas concentrations, there are two main ESI defined by the annual net primary production *annNPP* [KgC/m²/yr] and the total carbon stock *TCS* [KgC/m²/yr]. Both indicators showed a clear altitudinal pattern where higher values are located at lower altitudes (Figure 8-4b). The *annNPP* for tussocks was around 1.09 KgC/m²/yr in lower altitudes and decreased approximately 25% at mid altitudes and 40% at higher altitudes with an *annNPP* around 0.66 KgC/m²/yr. The *annNPP* for acaulescent rosettes from lower to higher elevations ranged from 0.69 to 0.55 KgC/m²/yr and for cushions from 0.64 to 0.57 KgC/m²/yr. The *TCS* showed a highly significant concentration in lower altitudes for tussocks 33.61 KgC/m²/yr, acaulescent rosettes 30.23 KgC/m²/yr and cushions 28.22 KgC/m²/yr; those carbon stocks reduced around 25% at mid elevations and up to 75% in higher elevations.

8.3.2.3 Regional climate regulation

The only ESI indicator identified to quantify the micro and regional climate regulation was the energy absorption by evapotranspiration, evaluated by the ESI *annET* [mm/yr] indicator, that show larger variations among all growth forms of vegetation and altitudinal ranges (Figure 8-4c). The ET has three main components, which contributed in average up to 2% from canopy evaporation, 31% from leaf transpiration and 67% from soil evaporation. The percentages varied based on the growth forms of vegetation, particularly for canopy transpiration which is only around 1% for acaulescent rosettes and cushions and for leaf transpiration which can vary up to 5% being larger for tussocks than the other growth forms of vegetation.

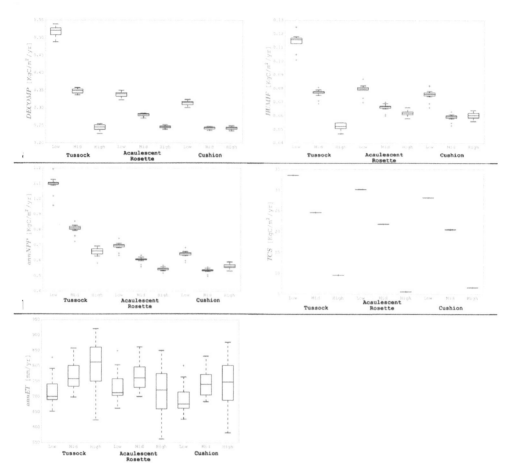

Figure 8-4 Ecosystem Service Indicators for a) Decomposition and fixing processes, b) Global climate regulation, and c) Regional climate regulation in the *páramos*.

8.3.3 Spatial distribution of ecosystem services

The spatial distribution of ecosystem service scores obtained for water regulation and carbon sequestration in the *páramo* ecosystem showed differences along an altitudinal gradient as evidenced earlier in the descriptive statistics (Figure 8-3 & Figure 8-4). Figure 8-5a showed the scores for the water regulation service, where high scores between 0.5 to 0.8 were located in areas dominated by tussocks at higher elevations and decreased around 0.3 to 0.4 for low and mid elevations. Conversely, acaulescent rosettes and cushions showed scores < 0.3 throughout the catchment area. For carbon sequestration the scores showed an altitudinal gradient where scores between 0.8 to 1.0 were found at lower altitudinal ranges and decreased to < 0.3 toward higher elevations.

Water regulation and carbon sequestration as the most significant human benefits identified in the *páramo* ecosystem are considered both equally important; however usually they are economically evaluated differently and therefore they should remain separately for further analysis.

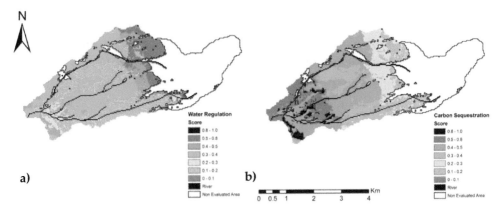

Figure 8-5 Averaged score of the benefits from ecosystem services for a) Water regulation and b) Carbon sequestration.

8.4 Discussion

8.4.1 Water regulation

8.4.1.1 Mass stabilization and control of erosion rates

All vegetated areas provide mass stabilization and protect the soil against erosion; however their growth forms of vegetation differ on how the erosion rates are controlled. Notably, tussocks stood out demonstrating a higher capacity in comparison to acaulescent rosettes and cushions. Tussocks have the highest amount of biomass protecting the soil against erosion which in turn provides better conditions for the regulation of flows. Tussocks have roots around 1.5 m long in average, which usually help to stabilize slopes in very steep areas like in these high-altitudinal ecosystems (Minaya et al., 2015a). Conversely, acaulescent rosettes have thick-tap roots of 0.15 m long, while cushions can have roots around 1 m but it depends on the species and soil texture (Attenborough, 1995). The particular trend where all three growth forms of vegetation provide higher soil stabilization at higher elevations might be attributed to the high concentrations of lignin in roots, which gives strength to the cell walls and litter, and is essential for water transport inside the plant as evidenced in earlier studies (Minaya et al., 2015a). The harsh environment of low oxygen concentration, high radiation, strong wind abrasion and night frost (Baruch, 1984) at elevations above 4400 m a.s.l. (High catchment) might have influenced the physiology of the plants to develop characteristics for adaptation and to give the soil stabilization needed to provide an ecological service.

8.4.1.2 Water flow maintenance

The ecosystem water outflow depends on the precipitation and the water in the soil that is lost to runoff and groundwater, which in turn depends on the soil storage capacity. The ability of damping the water outflow is highly related to the properties associated with the soil such as saturated water capacity and field capacity and therefore linked directly to the runoff generation mechanisms. This indicator requires a bit more of attention since the idea was to do a fine-tunning of the soil moisture with the hydrological conceptual model as discussed in Chapter 7. Unfortunately, the hydrological processes in the soil are still not well-understood and therefore the the water content in the soil is still not accurately quantified.

The main characteristic of the soils in the *páramo* vegetation are the high water-holding capacity and therefore high water retention showing great potential to act as a sponge intercepting the rainfall and releasing the water later during dry periods (Buytaert and Beven, 2011, 2009; Buytaert et al., 2010; Janeau et al., 2015). The latter is particularly true for tussocks. Whereas, cushions are located in areas with a high water table that are constantly wet and therefore their already saturated soil gives very low capacity to store additional amount of water before it overflows. Moreover, the *páramo* ecosystem is located in irregular topography with many less steep zones that can be hydrologically disconnected and yet providing extra buffer areas to dissipate the peak flow at the outlet of the catchment.

8.4.2 *Carbon sequestration*

8.4.2.1 Decomposition and fixing processes

During the regulation of soils, the organisms interact with each other to produce and decompose the organic compounds and biomass. The *páramo* vegetation has the capacity to transport water from the soil into the atmosphere and vice versa and by performing these functions, these terrestrial plants contribute to the regulation of soil, air and water quality in the region (Alcamo, 2003). These processes were higher at lower altitudinal elevations due to the higher availability of nutrients in the soils and higher temperatures, which enhance the soil microbial activity (O'Connor et al., 1999). The litter also plays a crucial role in the soil organic matter by transferring energy input for microflora and fauna (Bernhard-Reversat and Loumeto, 2002) and also for local atmospheric composition due to the CO_2 produced by microbial activity (Vanderbilt et al., 2008).

8.4.2.2 Global climate regulation

The slow decomposition rates enhance the carbon storage allowing a large amount of terrestrial carbon to be stored in the *páramo* soils (Farley et al., 2013) particularly in low elevations where the effective soil is deeper. Moreover, the carbon concentration in the plant tissues and soil are higher at low elevations in comparison to higher altitudes (Minaya et al., 2015a). The physicochemical quality and composition of the plant litter influences the ecosystem functioning, which in turn increase the amount of carbon in the soil thus important for global carbon cycle (Bell and Worrall, 2009). Although, most of the carbon stocks are sequestrated in soils, tussocks are distinctive due to their bulky above ground biomass notably at lower elevations. At higher elevations, the vegetation is more scattered, effective soil is less deep and therefore the carbon sequestration decreases considerably.

8.4.2.3 Regional climate regulation

The evapotranspiration is a broad term that comprises three main evaporative processes, and if possible should be treated separately (Savanije, 2004). Although interception is higher in tussocks compared to the other growth forms of vegetation, the evaporation from the canopy contributed in a very small percentage to the total evapotranspiration. During photosynthesis the plants take the CO_2 from the atmosphere and in turn release water as vapour through their stomata in a process known as transpiration and estimated as one third of the total evapotranspiration. In the *páramo* ecosystem, this process is limited by the fog and mist formed from the orographic uplift by the Andes mountains (Buytaert and Beven, 2011). Additionally, the *páramo* vegetation is well known to condensate the humidity of the air and transport that extra input of water into the system and therefore it slightly changes the water balance equation. Unfortunately, the latter was not considered into the study.

8.4.3 Implications for Payment for Ecosystem Services

The map of scores revealed spatially different capabilities of the landscape to provide services of water regulation and carbon sequestration along an altitudinal gradient and for the main growth forms of vegetation. Here, we do not imply that areas are more or less important. Instead, we quantify accurately the services that a *páramo* ecosystem within an ecological reserve area is providing so it can be used as a benchmark. The water regulation and carbon sequestration services often go hand in hand as they are not two completely independent services. Vegetation cover protects the soil against erosion, acts as buffer initially intercepting the incoming precipitation, reducing the peak flow level and thus providing water regulation. It also prevents a wash out of nutrients that are essential for vegetation growth, which in turn sequestrate carbon in their plant components and principally in soil in these *páramo* ecosystems.

The assessment presented in this study could potentially be used as an evaluation tool for quantification and spatial modelling of multiple ecosystem services at different scales in heterogeneous ecosystems such as the *páramo*, aiming at a fair payment for ecosystem services. The government initiatives to conserve and protect the *páramo* ecosystem have been extensively discussed in other studies (Farley et al., 2011; Farley et al., 2013). SocioParamo is a governmental initiative that was born from the importance to conserve the *páramo* ecosystem as a key source of water supply and it has had a good acceptance within the many communities in the zone (Farley et al., 2011; MAE., 2009). However, it is not based on any thematic nor spatial resolution and it contains only qualitative assessment with a high degree of generalization and a payment structure of 30 USD/ha/yr. The payment currently established is very low compared with what the international markets are currently paying (average price of the carbon sequestrated around 19 € TmC/yr).

Therefore, we believe that the improvement of the current methodology for assessing a payment for ecosystem services together with an alternative for livelihood strategy development could effectively engage a massive participation from land holders and communities to enrol in their programs. Nowadays, communities and other beneficiaries look at the *páramo* as useless land; mainly due to the lack of knowledge about its ecological services. Additionally, they have developed a high-altitude agriculture, where potato cultivation is a priority as the most profitable sowing, while livestock is not well organized and in a smaller scale (Zabala and Falconi, 2010).

Finally, the ecosystem biodiversity is an aspect that has been overlooked when it comes to the integral valorisation of ecosystem services. The *páramos* hold a great amount of biodiversity that includes a high number of species currently at risk and for that reason it adds environmental value (Castellanos, 2011; Downer, 1996; Naveda-Rodríguez et al., 2016; Sierra et al., 2002; Zapata-Ríos and Lyn, 2016) that should be considered in the valorisation of these key *páramo* ecosystems.

8.5 Conclusions

The main ecosystem services for the *páramo* ecosystem are water regulation and carbon sequestration and they are best quantified by looking at the different main growth forms of vegetation and altitudinal ranges.

For water regulation, notably the tussocks stood out demonstrating a higher capacity and providing better conditions for the regulation of flows throughout the catchment. Tussocks hold not only a high amount of biomass protecting the soil against erosion but also a high water-holding capacity in the soil which buffers the high peaks of precipitation and later release as a constant flow to the main streams. Whereas, cushions are located in areas that are constantly wet and therefore hold saturated soil that gives a small capacity to store additional amount of water before it overflows.

For carbon sequestration, the altitude and the effective soil depth contributed significantly. The decomposition and fixing processes were higher at lower altitudinal elevations due to the higher availability of nutrients in the soils and higher temperatures, which enhance the soil microbial activity. In general, the *páramo* vegetation is characterized by the slow decomposition rates allowing a large amount of terrestrial carbon to be stored in the *páramo* soils particularly in low elevations where the effective soil is deeper. At higher elevations, the vegetation is more scattered, effective soil is less deep and therefore the carbon sequestration decreases considerably.

All vegetated areas have a significant contribution in terms of water regulation and carbon sequestration and the overall analysis will contribute to improve the value and spatial distribution of ecosystem services in *páramo* regions. Since the case study lies in an Ecological Reserve, the quantification of ecosystem services at different altitudinal ranges can serve as a reference point for further monitoring other *páramo* sites that might be threatened by burning, grazing and land conversion. The results can be used as an integrated information approach to complement ongoing environmental management and landscape planning strategies (e.g. such as the program SocioPáramo) so that it can better meet the social goals. The graphical results could be use as a monitoring and discussion tool with the community in order to make easier capacity building, exchange of experiences and ideas, and concepts for coordinated actions among all stakeholders.

Further research can be the quantification of economic values of the human benefits to give a better assessment and support to the Payment for Ecosystem Services in public and private-sector decision making within the *páramo* ecosystems. The ultimate goal is to develop a *páramo* carbon program that might be the first of its kind in Ecuador under the Clean Development Mechanism of the Kyoto's Protocol (UNFCCC, 2008) in order to generate carbon credit financing that will support more of these social initiatives to benefit local communities.

You cannot teach a man anything; you can only help him to discover it in himself (Galileo)

CONCLUSIONS AND RECOMMENDATIONS

9.1 General

From all the ecosystems around the globe, tropical grasslands are the most abundant but probably the least understood in terms of their eco-hydrological processes, particularly the grasslands located in high-altitude regions. In the Andes, these ecosystems are known as *páramos*, which for many years have been recognized for the valuable ecosystem services they provide. However, their functioning has neither appropriately been assessed nor quantified. The IPCC 2013 Guidelines (IPCC, 2013) highlight the importance of quantifying above- and belowground carbon stocks and emissions from grasslands due to burning and land-use change. However, such recommendations do not include the influence of local factors such as altitude, vegetation and soil texture on carbon stocks, which are necessary in order to understand the complex biogeochemical interactions between soil and plant community. Ecuador has nearly 12500 km² of *páramo* from which more than 60% has degraded while the remaining areas are currently under constant pressure of land conversion. More than 80% of the drinking water for Quito, Ecuador's capital city, comes directly from the *páramos*. The lack of information about the spatial variation and temporal dimension of the main ecological processes has compromised their main ecosystem services, notably hydrological regulation and carbon sequestration. There are considerable gaps since almost no scientific studies have looked at the ecological processes and interactions in an integrative way, to consider the spatio-altitudinal variation of vegetation and soil characteristics as well as runoff generation processes of this complex ecosystem. An adequate *páramo* ecosystem management plan requires to first understand confidently the current processes and interactions of the *páramo* ecosystem before providing scientific tools to properly assess and quantify the ecosystem services to support policy mechanisms that can be broadly used by policy makers currently involved in the development of management strategies in the *páramos*. This is consistently promoted in the strategy suggested by the European Union to reinforce the sustainable development that aims to improve the well-being of the population as well as the conservation, restoration and sustainable use of strategic ecosystems.

This research contributes to the understanding of the interactions and functioning of the *páramo* ecosystem with a case study in northern Ecuador as a step towards an effective ecosystem management plan. In addition, one of the main priorities in Chapter 4 of the IPCC Report is 'to improve representation of the interactive coupling between ecosystems and the climate system'. Certainly, this can be achieved by reducing the uncertainties in GPP estimations and threshold responses by using more realistic characteristics of the ecosystems.

This study can be used as a tool to check the sustainable use of the *páramos* by exploring potential effects such as intensive grazing, land conversion and burning.

9.2 Main contributions

> *Assessment of carbon and nitrogen concentrations in soil and vegetation, aboveground carbon stocks distribution and soil organic carbon stocks at different altitudinal ranges*

Several studies looking at climate change dynamics using ecophysiological information of terrestrial ecosystems ended up using generalized parameterization due to the lack of information. This was evident during literature review, we came across that there was not readily available information of ecophysiological parameters for *páramo* grasslands, and even the little information available for grasses did not discriminate between C3 and C4 classes, which use different enzymes to accept CO_2 from the atmosphere. In this regard, a key contribution of this research is the altitudinal analysis that supported a potential distribution of carbon and nitrogen concentrations in soil, litter and live tissues, where higher concentrations were found in the low altitudinal ranges mainly for tussocks and acaulescent rosettes. Soil texture provided complementary information: a high percentage of silt was highly correlated to high soil nitrogen and carbon concentration.

The analysis of the carbon and nitrogen concentrations as well as other carbon pools parameter selection provide significantly to the documentation for critical parameters that often are limiting factors for accuracy and credibility of regional and global model of carbon fluxes simulations. We believe that authentic estimations of the spatial distribution of key parameters would contribute to a considerable reduction in the uncertainty of gross and net primary production simulations. Although, these estimations were measured at our study area, the values can entirely be used for similar growth forms in neighbouring catchments as well as other *páramo* ecosystems in the Andes. In Chapter 3, the critical parameters are well documented with a mean and standard deviation as our initial measure of variability. The changes in the ecological and physiological properties have a direct impact on the water cycle and primary productions of these high-altitudinal ecosystems.

Implementation of selected biogeochemical and ecophysiological process models to simulate carbon and water fluxes in the páramo ecosystem; testing model performance, in particular on gross primary production and water budget in the páramo system

Building an eco-hydrological model for the *páramo* ecosystem required several steps, which include the selection of an appropriate model capable to deal with the complexity of the high spatial variability and heterogeneity of the site as well as the representation of the most relevant physical processes and/or the possibility to adapt to our needs without significant structural changes. Among the many ecophysiological and biochemical models, the BIOME-BGC was chosen for its suitability and it was used further in this research.

The ecosystem process-based model BIOME-BGC of the University of Montana was used at three different elevations taking into account the distinctive physiology of three main growth forms of *páramo* vegetation. Given that the main purpose is to simulate the carbon and water fluxes, or more specifically the gross primary production (GPP), which had proved to be a good indicator of ecosystem's health and the soil moisture and water storage as key elements of this important ecosystem. It was very difficult to validate the simulation of the BIOME-BGC due to the inexistence of in-situ GPP measurements as well as the lack of soil moisture measured data. However, the parameterization and calibration processes rely on the statistical analysis of key parameters derived from in-situ measurements in order to produce a significant reduction in the uncertainty of the simulations. The main contribution of Chapter 4 is the detailed analysis of GPP at different altitudinal ranges and the relationship between each growth form to the total GPP. It showed that the interaction between vegetation surface and the atmosphere such as radiation uptake, precipitation interception, energy conversion, momentum and flux exchange are extensively determined by the vegetation surface in the *páramos*. Unfortunately, the BIOME-BGC has limitations in the conceptualization of the soil component; it uses a simplified approach of a 1D bucket model that limit the possibility for the model to capture the soil profile dynamics and soil water capacity.

Analysis of the relationship between climatic variables and gross primary production using data–driven modelling techniques

After testing the ecophysiological and biogeochemical model BIOME-BGC along an altitudinal gradient, a number of statistical data analysis and data driven models were used to evaluate in a rapid way the complex relationship between GPP and diagnostic variables (Temperature, Precipitation, Short wave radiation, vapour pressure deficit) at various time frames. For this, a two-step approach was used, a model-free and then a model-based technique to analyse the performance of linear regression model, model tree, instance-based learning and artificial neural network model to simulate temporal variations of GPP at three altitudinal ranges. We found that the relationship between GPP and the set of input climatic variables was stronger with different time aggregation; there is added value in computing the GPP based on meteorological parameters that are easier to measure and are readily available as climate time-series for future scenarios. The results of bi-monthly and monthly time frames were similar and reasonably good. On the other hand, the short computational time required for running the DDMs allows a faster extrapolation to higher spatio-temporal scales, which can be an effective approach especially when using climate change scenarios such as the CMIP (Coupled Model Intercomparison Project).

It is worth to mention that these surrogate models do not replace a detailed and comprehensive physical based model; instead the analysis showed that there is good potential of using the DDMs are complementary statistical technique for assessing the ecosystem based only on determined input set of variables. These results emphasize the importance to count on reliable observation-based data sets to correctly estimate GPP and better quantify the future uptake of CO_2 by the *páramo* vegetation. Accurate estimations of GPP can have important consequence on other terrestrial fluxes (e.g. NPP, NEP, energy/water fluxes) and reservoirs (e.g. soil carbon stocks).

Determining the origin and quantifying the contributions of the main runoff components using environmental tracers (isotopes and major ions)

An extensive fieldwork campaign was carried out to identify the most dominant processes and temporal contribution of the runoff components, such as glacier, rainfall, surface runoff, and subsurface flow, to the total runoff. Based on this, Chapter 6 showed a spatial hydrochemical analysis and the suitability of tracers that were able to distinctively identify groups by source, by geographical location and geological background. A comprehensive analysis of surface water determined two different components being: *páramo* and glacier, which were quantified during dry conditions and rainfall events. The main contribution of this chapter lies in the fact that it provides a full spatial distribution of the hydrochemical characteristics of the streams per subcatchment, identifying possible runoff generation processes along the streams. The hydrograph separation showed that the contribution of the glacier component during a rainfall event increases at a faster rate than the *páramo*, which behave as a floodplains and wetlands that dissipate the stream energy and buffer the peak flow.

The temporal resolution was beyond the scope of this PhD research (only one field study was carried out which already required considerable effort) and certainly this needs to be further studied because of the fact that the behaviour of the streams from glaciers depend on the glacier influence, for example, its melting during diurnal cycles are different from the nocturnal cycles. However, we identified a couple of sources where there are clear evidence of resurgence of the meltwater from the glacier consistently characterized with low values of EC. The results are of great importance for the understanding of runoff generation mechanisms in this combined glacier and *páramo* catchment in the Ecuadorian Andean Region and it contributes to the limited number of hydrological studies specifically hydrograph separation, and quantification of different components. Certainly, long-term analysis will contribute to a better understanding on the dependency of runoff generation on soil moisture and vegetation interaction.

Applying a process–oriented hydrological model that represents the different runoff generation processes within the catchment

---

Runoff generation and hydrological processes in the *páramo* soils were poorly understood as evidence in other few attempts to model the *páramo* hydrology. This was highlighted as one of the many scientific challenges in the effort to understand the *páramo* hydrology (Buytaert et al., 2006a). In this regard, we selected a model that deal with the complexity of the runoff process, which is a challenging issue in high-altitudinal catchments. The model selected was the Tracer Aided Catchment model (TAC^D), which is able to describe the lateral water fluxes based on the soil properties, land use and topography, the governing equation for lateral fluxes is the simple differential equation of a linear storage unit and for vertical fluxes are represented by constant amounts of percolating water. The TAC^D described in several hydrological unit types how the surface water flux is routed through two or three storages interconnected via vertical or lateral fluxes. Unfortunately, the main objective of Chapter 7 was not met in the sense that the hydrological processes could not be explained fully and therefore the runoff generation processes in the hydrological system are still not well-understood. The level of representation of the hydrological processes within the model did not allow us to achieve a fair understanding of the variation of soil moisture since the spatio-temporal representation was too complex and difficult to validate. Time constrain did not allow us to refine the hydrological processes conceptualized for each hydrological unit. The main issue with this type of conceptualized models is the equifinality, which is a problem of the model and not necessarily with the calibration.

Further analysis should be done in the parameterization of the soil and runoff generation routine of the TAC^D as well as a better conceptualization of the runoff generation processes of each hydrological unit.

Leaving aside the data uncertainty due to the lack of reliable long time-series of climate data, which is not within the scope of this research, we critically evaluate the model conceptualization and parameterization as well as the simplification in the model structure in order to improve the potential use of TAC^D model in larger spatial scales without increasing its already complexity and computational time cost.

Assessing the ecosystem services of the páramos based on key indicators of regulation and maintenance

---

Páramos have been acknowledged as key components that provide important ecological services such as regulation & maintenance including biodiversity, scenic landscape, carbon sequestration and water regulation as extensively found in literature; however rarely quantified. The *páramo* ecosystem is currently facing environmental and social challenges as a result of a constant pressure due to land conversion, cutting and clearing for potato cultivation and grazing. This pressure is highly associated to socio-economic factors from communities that aspire for a higher income generation. Several national initiatives have been taking place in order to compensate for the ecosystem services provided; however those are not based on a solid scientific basis for negotiation. The current monetary amount given is per Ha without any assessment of scientific studies to support that payment.

In this regard, the contribution of the last chapter is an ecosystem service assessment that uses information from all previous integrated results of carbon stocks and water resources availability in the region. The ecosystem services indicators derived in this chapter accurately quantified the two most important services provided by the *páramo* being water regulation and carbon sequestration. Outcomes are descriptive statistics and maps that graphically depicted the potentials of certain areas to provide higher ecosystem services, which can potentially be used for the dissemination, monitoring and evaluation of the ecosystem management strategies with the local community and other key stakeholders. Although an economic valuation of these services is not within the scope of the present study, this analysis will definitely complement other current initiatives such as payment for ecosystem services with an adequate assessment for a fair valuation.

It is worth mentioning that the research is aimed to achieve part of the targets of the Sustainable Development Goals (SDGs) specifically #6 and #15 (UN, 2015). SDG 6: Clean water and sanitation *'By 2020, protect and restore water-related ecosystems, including mountains, forests, wetlands, rivers, aquifers and lakes'*. SDG 15: Life on Land *'By 2020, ensure the conservation, restoration and sustainable use of terrestrial and inland freshwater ecosystems and their services'* and *'By 2020, integrate ecosystem and biodiversity values into national and local planning, development processes, poverty reduction strategies and accounts'*.

9.3 Recommendations for future research

This research integrated all the different aspects of the *páramo* ecosystems, high spatial and altitudinal variations of climactic data and ecophysiological characteristics of the main predominant vegetation. However, the approach developed contained limitations and remains with several challenges that should be further addressed in order to improve reliability on the eco-hydrological simulations and ecosystem services quantification.

Further improvement in the spatial variability and interpolation of climatic drivers

The most important meteorological parameter that characterizes the rain gauging stations in the Ecuadorian Cordillera is the wind, which separates two periods, the first characterized by strong and almost constant winds from the east (between April and September) and the second period with weak and intermittent wind (between October and January) (Francou et al., 2004). The area of the Antisana is representative for the eastern mountainous region and it is directly exposed to the humid wind of the Amazon River Basin (Manciati et al., 2011). Unfortunately, we did not have reliable data of wind speed and direction that can correct the high spatial variability of precipitation in our study area. We performed several comparisons among different approaches to interpolate precipitation in the catchment; the technique chosen was the inverse distance interpolation and the inclusion of elevation weighing since we considered that elevation could play a major role in the distribution of the precipitation in these mountainous regions. Precipitation as input data is very important for the hydrological modelling and therefore the method used should integrate wind as a key factor for interpolation within these mountainous regions with high spatial variability. Additionally, there was no data of the so called "horizontal precipitation", which is known to add some water into the system. In several hydrological studies in the area, this term has been mentioned but so far neglected (Buytaert et al., 2006a; Buytaert et al., 2005b). Studies in the Colombian *páramos* (Tobon and Gil Morales, 2007) have demonstrated that typical *páramo* vegetation can intercept in average 0.22 mm/h, this value depends on both the vegetation cover and the density of the fog event. Therefore, we consider crucial to measure its intensity, duration, and frequency regime at different altitudinal ranges and then its results can later included as an additional input data in all hydrological conceptualizations in these types of ecosystems.

The *páramo* ecosystems are characterized by a low evaporation and high saturation vapour and therefore generating a high hydrological efficiency in the catchment (Tobon and Gil Morales, 2007). The term evapotranspiration should be taken wisely and every component carefully checked. For our case we believe all elements are equally important and exceptionally distinctive in the *páramo* ecosystem. In this regard, we look in detail the estimation of interception, evaporation and transpiration. However, there is still room for further improvement in the evaluation of these critical parameters such as flux measurements of eddy covariance and porometry techniques in-situ. Observed values of latent, sensible heat flux and water vapour loss from a leaf can help to evaluate the reliability of our parameters. The incorrect estimation of evapotranspiration and all its components lead to a wrong estimation of other terms of the hydrological balance such as soil moisture and stream flows (when discharge data is not available).

Additional developments in the process model structure

The BIOME BGC presented several limitations, particularly for herbaceous vegetation as evidenced during our simulations, where the soil moisture was not correctly estimated due to the simple soil hydrology within the model structure. The *páramo* vegetation is sensitive to soil processes and therefore it needs an appropriate soil module that can handle the processes of runoff, diffusion and percolation and that additionally considers the stress on plant mortality.

Although the process-oriented hydrological model TAC^D was tested in other high mountainous catchments, it was not suitable for the application in the *páramo* ecosystem due to inaccurate simulation of the discharge at the outlet of the basin. The model was flexible to incorporate findings where runoff is generated differently taking into account topographical, morphological and ecological features. However, it still needs further improvements in the hydrological processes conceptualizations of each of the hydrological units in order to demonstrate plausibility in the application to high-altitudinal *páramo* ecosystems. We are aware that highly parameterized models yield to high uncertainties that inhibit inference on the model simulations. For example, it would be crucial to perform a sensitivity analysis of the parameters values to give further understanding of the internal model behaviour. It is well known that there are other types of uncertainty to be considered, such as uncertainty of the input data and model structure. The latter should be further analysed to guarantee the usefulness of the model to estimate soil moisture and all hydrological processes within these high-altitudinal catchments.

Enhancing the potential use of Data-Driven Modelling techniques

It was beyond this PhD research to test the potential use of DDM with available data sets to estimate GPP in other Andean regions. However, we see that the research might contribute to a good assessment of the current GPP estimations in other *páramo* regions in the Andes, since the use of classical models for a high spatial resolution limit the assessment capacity due to the expensive computational time. The use of these techniques could be further enhanced to evaluate the possibility to complement the use of process-based models and therefore supporting the prediction of future scenarios.

Comprehensive assessment of Ecosystem services

Water managers, scientific and local community as direct actors for decision making always look for good assessment strategies that can help to evaluate the ecosystem capacities to provide services in a spatial manner. In this regard, it will be crucial to enhance the present study with an economic valuation for a fair payment of the ecosystem services to the community. It is important to look for a mature and more integrative approach in which cultural services are also included; however, this remains a challenge due to its complex and subjective way of estimation. We believe that the amount paid for the ecosystem services should meet the opportunity cost to keep protecting the current *páramo* and recover those areas that went under land conversion. The benefits could be a competitive opportunity to the current "profitable" potato cultivation. The ultimate goal is to develop a *páramo* carbon program that might be the first of its kind in Ecuador under the Clean Development

Mechanism of the Kyoto's Protocol in order to generate carbon credit financing that will support more of these social initiatives to benefit local communities.

10

REFERENCES

Aber, J. D.: Why don't we believe the models. , Bull. Ecol. Soc. Am., 232–233, 1997.

ADF: Fiber (Acid Detergent) and Lignin in Animal Feed (973.18). In: Official Methods of Analysis, Association of Official Analytical Chemists. 15th Edition. (Modifications: Whatman 934-AH glass micro-fiber filters with 1.5um particle retention used in place of fritted glass crucible.) ADF method only., 1990.

Aha, D., Kibler, D., and Albert, M.: Instance-based learning algorithms, Machine Learning, 6, 37-66, 1991.

Alcamo, J., et al. (Ed.): Ecosystems and human well-being : a framework for assessment, ISLAND PRESS, Washington, 2003.

Alvarez, R. and Lavado, R.: Climate, organic matter and clay content relationships in the Pampa and Chaco soils, Argentina Geoderma, 83, 127–141, 1998.

Anav, A., Friedlingstein, P., Beer, C., Ciais, P., Harper, A., Jones, C., Murray-Tortarolo, G., Papale, D., Parazoo, N. C., Peylin, P., Piao, S., Sitch, S., Viovy, N., Wiltshire, A., and Zhao, M.: Spatiotemporal patterns of terrestrial gross primary production: A review, Review of Geophysics, 53, 785-818, 2015.

Anderson, M. J.: Distance-based tests for homogeneity of multivariate dispersions, Biometrics, 62, 245-253, 2006.

Anderson, M. J.: A new method for non-parametric multivariate analysis of variance, Austral Ecology, 26, 32-46, 2001.

Anthelme, F. and Dangles, O.: Plant-plant interactions in tropical alpine environments, Perspectives in Plant Ecology, Evolution and Systematics, 14, 363-372, 2012.

Armijos, M. T. and De Bièvre, S.: El Páramo como proveedor de servicio ambiental primordial, El Agua. Contribución al estado del conocimiento y conservación de los Páramos Andinos In: Libro de Investigación Del Proyecto Páramo Andino, Condesan, Universidad de Amsterdam, Universidad de Wisconsin, Quito, 2012.

Arnell, N. W.: Climate change and global water resources, Global Environmental Change, 31–49, 1999.

Arneth, A., Miller, P. A., Scholze, M., Hickler, T., Schurgers, G., Smith, B., and Prentice, I. C.: CO_2 inhibition of global terrestrial isoprene emissions: Potential implications for atmospheric chemistry Geophys. Res. Lett., 34, 2007a.

Arneth, A., Niinemets, U., Pressley, S., Back, J., Hari, P., Karl, T., Noe, S., Prentice, I. C., Serca, D., Hickler, T., Wolf, A., and Smith, B.: Process-based estimates of terrestrial ecosys-tem isoprene emissions: incorporating the effects of a direct CO_2 isoprene interaction Atmos. Chem. Phys., 7, 31–53, 2007b.

Ataroff, M. and Rada, F.: Deforestation impact on water dynamics in a Venezuelan Andean cloud forest Ambio, 29, 440–444, 2000.

Attenborough, D.: The Private Life of Plants: A Natural History of Plant Behavior, BBC Books, London, 1995.

Azocar, A. and Rada, F.: Ecofisiología de plantas de páramo. Instituto de Ciencias Ambientales (ICAE), Facultad de Ciencias. Universidad de los Andes Mérida, Venezuela, 182 p pp., 2006.

Bagstad, K. J., Semmens, D. J., Waage, S., and Winthrop, R.: A comparative assessment of desicion-support tools for ecosystem services quantification and valuation, Ecosystem Services, 5, 27-39, 2013.

Band, L. E., Patterson, P., Nemani, R., and Running, S. W.: Forest ecosystem processes at the watershed scale: incorporating hillslope hydrology. , Agric. For. Meteorol., 63, 93-126, 1993.

Baruch, Z.: Ordination and Classification of Vegetation along an Altitudinal Gradient in the Venezuelan Páramos, Vegetation, 2, 115–126, 1984.

Beckers, J. and Alila, Y.: A model of rapid preferential hillslope runoff contributions to peak flow generation in a temperate rain forest watershed Water Resources Research 40, 2004.

Beckers, J., Smerdon, B., Redding, T., Anderson, A., Pike, R., and Werner, A. T.: Hydrologic Models for Forest Management Applications: Part 1: Model Selection, Watershed Management Bulletin 13, 2009a.

Beckers, J., Smerdon, B., and Wilson, M.: Review of hydrologic models for forest management and climate change applications in British Columbia and Alberta. Forrex Series 25, 2009b.

Beer, C., Reichstein, M., Tomelleri, E., Ciais, P., Jung, M., Carvalhais, N., Rödenbeck, C., Arain, A., Baldocchi, D., Bonan, G. B., Bondeau, A., Cescatti, A., Lasslop, G., Lindroth, A., Lomas, M., Luyssaert, S., Margolis, H., Oleson, K. W., Roupsard, O., Veenendaal, E., Viovy, N., Williams, C., Woodward, I., and Papale, D.: Terrestrial Gross Carbon Dioxide Uptake: Global Distribution and Covariation with Climate, Science, 329, 834-838, 2010.

Belgrano, A., Malmgren, B. A., and Lindahl, O.: Application of artificial neural networks (ANN) to primary production time-series data, Journal of Plankton Research, 23, 651-658, 2001.

Bell, M. J. and Worrall, F.: Estimating a Region's Soil Organic Carbon Baseline: The Undervalued Role of Land-Management. , Geoderma, 152, 74–84, 2009.

Benavides, J.: The Effect of Drainage on Organic Matter Accumulation and Plant Communities of High-Altitude Peatlands in the Colombian Tropical Andes Mires and Peat, 15, 1–15, 2014.

Beniston, M.: Climatic change in mountain regions: A review of possible impacts, Clim. Change, 59, 5-31, 2003.

Bergstrom, S.: The HBV model, Chapter 13 of Computer models of watershed hydrology Water Resour. Publications, 443-476, 1995.

Bernhard-Reversat, F. and Loumeto, J. J. (Eds.): The litter system in African forest tree plantations Science Publishers, Inc., Plymouth, UK, 2002.

Beven, K.: A manifesto for the equifinality thesis Journal of Hydrology, 320, 18-36, 2006.

BioVel Portal: http://ecos.okologia.mta.hu/bbgcdb/, last access: June 10th 2016.

Blum, A. L. and Langley, P.: Selection of relevant features and examples in machine learning, Artif. Intell., 97, 245–271, 1997.

Bohn, H. G.: Humboldt, Alexander von. Personal Narrative of Travels to the Equinoctial Regions of America During the Years 1799-1804, Chapter 25. London, 1853.

Bond-Lamberty, B., Gower, S. T., Ahl, D. E., and Thornton, P. E.: Reimplementation of the Biome-BGC model to simulate successional change, Tree Physiology, 25, 413-424, 2005.

Bond-Lamberty, B., Peckham, S. D., Gower, S. T., and Ewers, B. E.: Effects of fire on regional evapotranspiration in the central Canadian boreal forest Global Change Biology, 15, 1242-1254, 2009.

Boogaard, H. L., De Wit, A. J. W., te Roller, J. A., and Van Diepen, C. A. (Eds.): User's guide for the WOFOST Control Center 1.8 and WOFOST 7.1.3 crop growth simulation model, Wageningen, 2011.

Boogaard, H. L., van Diepen, C. A., Rötter, R. P., Cabrera, J. M. C. A., and van Laar, H. H.: User's guide for the WOFOST 7.1 crop growth simulation model and WOFOST Control Center 1.5. Tech. Doc. 52. , Wageningen, the Netherlands., 1998.

Bosman, A. F., van der Molen, P. C., Young, R., and Cleef, A. M.: Ecology of a Paramo Cushion Mire Journal of Vegetation Science, 4, 633–640, 1993.

Bowling, L. C., Storck, P., and Lettenmaier, D. P.: Hydrologic effects of logging in western Washington, United States Water Resources Research 36, 3223 – 3240, 2000.

Brack Egg, A.: Las correcciones del Peru, Boletin de Lima 1986.

Bradley, R. S., Vuille, M., Diaz, H. F., and Vergara, W.: Threats to Water Supplies in the Tropical Andes, Science, 312, 1755-1756, 2006.

Braun-Blanquet., J.: Pflanzensoziologie.3rd ed., Berlin, Vienna, New York, Springer-Verlag, 1964.

Bristow, K. L. and Campbell, G. S.: On the relationship between incoming solar radiation and daily maximum and minimum temperature, Agricultural and Forest Meteorology, 31, 159-166, 1984.

Brown, D. G., Lusch, D. P., and Duda, K. A.: Supervised classification of types of glaciated landscapes using digital elevation data, Geomorphology, 21, 233-250, 1998.

Brown, L. E., Milner, A. M., and Hannah, D. M.: Predicting river ecosystem response to glacial meltwater dynamic: a case study of quantitative water sourcing and glaciality index approaches, Aquat. Sci, 72, 325-334, 2010.

Bruijnzeel, L. A.: Hydrological functions of tropical forests: Not seeing the soil for the trees? , Agriculture Ecosystems & Environment 104, 185–228, 2004.

Bruijnzeel, L. A. and Proctor, J. (Eds.): Hydrology and biogeochemistry of tropical montane cloud forests: What do we really know? In: Hamilton, L.S., Juvik, J.O., Scatena, F.N. (eds.), , Tropical montane cloud forests. Springer-Verlag, New York. pp. 38-78., 1995.

Buckee, G. K.: Determination of total nitrogen in barley, malt and beer by Kjeldahl procedures and the Dumas Combustion Method, J. Inst. Brewing, 100, 57-64, 1994.

Buttle, J. M.: Isotope hydrograph separations and rapid delivery of pre-event water from drainage basins Progress in Physical Geography, 18, 16-41, 1994.

Buytaert, W. and Beven, K.: Models as multiple working hypotheses: hydrological simulation of tropical alpine wetlands, Hydrological Processes, 25, 1784-1799, 2011.

Buytaert, W. and Beven, K.: Regionalization as a learning process, Water resources Research, 45,W11419, 2009.

Buytaert, W., Celleri, R., De Bievre, B., Hofstede, R., Cisneros, F., Wyseure, G., and Deckers, J.: Human impact on the hydrology of the Andean paramos, Earth Science Reviews 79, 53–72, 2006a.

Buytaert, W., Celleri, R., Willems, P., De Bievre, B., and Wyseure, G.: Spatial and temporal rainfall variability in mountainous areas: a case study from the south Ecuadorian Andes, journal of Hydrology, 329, 413-421, 2006b.

Buytaert, W., Cuesta-Camacho, F., and Tobón, C.: Potential impacts of climate change on the environmental services of humid tropical alpine regions, Global Ecology and Biogeography, 20, 19-33, 2011.

Buytaert, W., De Bievre, B., Wyseure, G., and Deckers, J.: The use of the linear reservoir concept to quantify the impact of land use changes on the hydrology of catchments in the Ecuadorian Andes Hydrology and Earth System Sciences 8, 108–114, 2004.

Buytaert, W., Deckers, J., and Wyseure, G.: Description and classification of nonallophanic Andosols in south Ecuadorian alpine grasslands (páramo), Geomorphology, 73, 207-221, 2005a.

Buytaert, W., Iniguez, V., Celleri, R., De Bievre, B., Wyseure, G., and Deckers, J.: Analysis of the water balance of small paramo catchments in south Ecuador., Bergen, Norway, 20–23 June.2005b.

Buytaert, W., Sevink, J., Leeuw, B. D., and Deckers, J.: Clay mineralogy of the soils in the south Ecuadorian paramo region Geoderma 127, 114–129, 2005c.

Buytaert, W., Vuille, M., Dewulf, A., Urrutia, R., Karmalkar, A., and Celleri, R.: Uncertainties in climate change projections and regional downscaling in the tropical Andes: implications for water resources management, Hydrology and Earth System Sciences, 14, 1247-1258, 2010.

Buytaert, W., Wyseure, G., De Bievre, B., and Deckers, J.: The effect of land-use changes on the hydrological behaviour of Histic Andosols in south Ecuador, Hydrological Processes, 19, 3985-3997, 2005d.

Cáceres, B., Maisincho, L., Taupin, J. D., Francou, B., Cadier, E., Delachaux, F., Bucher, R., Villacis, M., Paredes, D., Chazarin, J. P., Garces, A., and Laval, R.: Glaciares del Ecuador: Antizana y Carihauyrazo. Balance de masa, topografía, meteorología e hidrología, IRD/INAMHI/EMAAP, 171 pp., Quito, 2005.

Cadier, E., Villacis, M., Garces, A., Maisincho, L., Laval, R., Paredes, D., Caceres, B., and Francou, B.: Variations of a low latitude Andean glacier according to gloval and local climate variations: first results. , Glacier Mass Balance Changes and Meltwater Discharge. (Selected papers from sessions at the IAHS Assembly in Foz do Iguazy, Brazil, 2005), 66-74, 2007.

Cañadas Cruz, L.: El mapa bioclimático y ecológico del Ecuador, Banco Central del Ecuador, Quito, 1983.

Carrillo-Rojas, G., Silva, B., Córdova, M., Célleri, R., and Bendix, J.: Dynamic mapping of evapotranspiration using an energy balance-based model over an andean páramo catchment of Southern Ecuador, Remote Sensing, 8, 1-24, 2016.

Carvalhais, N., Reichstein, M., Seixas, J., Collatz, J. C., Pereira, J. S., Berbigier, P., Carrara, A., Granier, A., Montagnani, L., Papale, D., Rambal, S., Sanz, J. M., and Valentini, R.: Implications of the carbon cycle steady state assumption for biogeochemical modeling performance and inverse parameter retrieval, Global Biogeochem. Cycles 22, GB2007, doi:2010.1029/2007GB003033., 2008.

Castellanos, A.: Andean Bear Home Ranges in the Intag Region, Ecuador, Ursus, 22, 65–73, 2011.

Cauvy-Frauníe, S., Condom, T., Rabatel, A., Villacis, M., Jacobsen, D., and Dangles, O.: Techincal Note: Glacial influence in tropical mountain hydrosystems evidenced by the diurnal cycle in water levels, Hydrology and Earth System Sciences, 17, 4803-4816, 2013.

Cavalier, J. (Ed.): Environmental factors and ecophysiological processes along altitude gradients in wet tropical mountains. In Tropical forest Ecophysiology. , 1996.

Cavieres, L. A., Quiroz, C. L., Molina-Montenegro, M. A., Muñoz, A. A., and Pauchard, A.: Nurse effect of the native cushion plant *Azorella monantha* on the invasive non-native *Taraxacum officinale* in the high-Andes of central Chile, Perspectives in Plant Ecology, Evolution and Systematics, 7, 217-226, 2005.

Celleri, R. and Feyen, J.: The Hydrology of Tropical Andean Ecosystems: Importance, Knowledge Status, and Perspectives, International Mountain Society, 29, 350-355, 2009.

Celleri, R., Willems, P., Buytaert, W., and Feyen, J.: Spacetime variability of rainfall in the Paute basin, Ecuadorian Andes Hydrological Processes, 21, 3316–3327, 2007.

Chen, H.: Biotechnology of Lignocellulose: Theory and Practice, Springer Netherlands, Dordrecht, 2014.

Chen, Q. and Mynett, A. E.: Modelling algal blooms in the Dutch coastal waters by integrated numerical and fuzzy cellular automata approaches, pg 73(79) pp., 2006.

Chow, V. T.: Open-channel hydraulics McGraw-Hill, 680 p, New York, 1959.

Chow, V. T., Maidment, D. R., and Mays, L. W.: Applied Hydrology McGraw Hill, New York, 1988.

Christophersen, N. and Hooper, R. P.: Multivariate-Analysis of Stream Water Chemical-Data - the Use of Principal Components-Analysis for the End-Member Mixing Problem. , Water Resources Research, 28, 99-107, 1992.

Churkina, G., Tenhunen, J., Thornton, P., Falge, E. M., Elbers, J. A., Erhard, M., Grunwald, T., Kowalski, A. S., Rannik, U., and Sprinz, D.: Analyzing the Ecosystem Carbon Dynamics of Four European Coniferous Forests Using a Biogeochemistry Model, Ecosystems, 6, 168-184, 2003.

Churkina, G., Zaehle, S., Hughes, J., Viovy, N., Chen, Y., Jung, M., Heumann, B. W., Ramankutty, N., Rödenbeck, C., Heimann, M., and Jones, C. D.: Interactions between nitrogen deposition, land cover conversion, and climate change determine the contemporary carbon balance of Europe Biogeosciences, 7, 2749-2764, 2010.

Clark, D. A., Brown, S., Kicklighter, D. W., Chambers, J. Q., Thomlinson, J. R., and Ni, J.: Measuring net primary production in forests: concepts and field methods, Ecological Applications, 11, 356-370, 2001.

Cleef, A. M. (Ed.): Characteristics of neotropical paramo vegetation and its subantartic relations, Geoecological relations between the southern temperate zone and the tropical mountains, 1978.

Cleef, A. M.: The Vegetation of the Páramos of the Colombian Cordillera Oriental. Dissertationes Botanicæ 61. J. Cramer, Vaduz., 1981.

Cleveland, C. C., Townsend, A. R., Schimel, D. S., Fisher, H., Howarth, R. W., Hedin, L. O., Perakis, S. S., Latty, E. F., Von Fischer, J. C., Elseroad, A., and Wasson, M. F.: Global patterns of terrestrial biological nitrogen (N_2) fixation in natural ecosystems, Global Biogeochemical Cycles, 13, 623-645, 1999.

Condom, T., Escobar, M., Purkey, D., Pouget, J. C., Suarez, W., Ramos, C., Apaestegui, J., Tacsi, A., and Gomez, J.: Simulating the implications of glaciers' retreat for water management: a case study in the Rio Santa basin, Peru, Water Int, 37, 442-459, 2012.

Connant, R. T., Paustian, K., and Elliott, E. W.: Grassland Management and Conversion into Grassland: Effects on Soil Carbon, Ecological Applications, 11, 343-355, 2001.

Corzo, G. A., Solomatine, D. P., Hidayat, de Wit, M., Werner, M., Uhlenbrook, S., and Price, R. K.: Combining semi-distributed process-based and data-driven models in flow simulation: a case study of the Meuse river basin, Hydrology and Earth System Sciences, 13, 1619-1634, 2009.

Cosby, B. J., Hornberger, G. M., and et.al.: A Statistical Exploration of the Relationships of Soil Moisture Characteristics to the Physical Properties of Soils., Water Resour. Res, 20, 671-681, 1984.

Costanza, R., de Groot, R., Sutton, P., van der Ploeg, S., Anderson, S. J., Kubiszewski, I., Farber, S., and Turner, R. K.: Changes in the Global Value of Ecosystem Services Global Environmental Change 152–158. doi:110.1016/j.gloenvcha.2014.1004.1002, 2014.

Cramer, W., Bondeau, A., Woodward, F., Prentice, I. C., Betts, R. A., Brovkin, V., Cox, P. M., Fisher, V., Foley, J. A., Friend, A. D., Kucharik, C., Lomas, M. R., Ramankutty, N., Sitch, S., Smith, B., White, A., and Young-Molling, A.: Global response of terrestrial ecosystem structure and function to CO2 and climate change: Results from six dynamic global vegetation models, Glob Chang Biology, 7, 357-373, 2001.

Crockford, R. H. and Richardson, D. P.: Partitioning of rainfall into throughfall, stemflow and interception: effect of forest type, ground cover and climate, Hydrological Processes, 14, 2903-2920, 2000.

Cuatrecasas, J.: Aspectos de la vegetacion natural de Colombia, Revista de la Academica Colombiana de Ciencias Exactas y Fisicas, 10, 221-264, 1958.

Cuatrecasas, J. (Ed.): Growth forms of the Espeletiinae and their correlation to vegetation types of the high tropical Andes, Academic Press, London, 1979.

Cuatrecasas, J.: Páramo vegetation and its life forms, In: Troll C (ed). Colloquium Geographicum 9. Geo-ecology of the mountainous regions of the tropical Americas. Ferdinand Dümmler, Bonn, pp163-186, 1968.

Cuesta, F., Báez, S., Muriel, P., and Salgado, S.: La vegetación de los Páramos del Ecuador. Contribución al estado del conocimiento y conservación de los Páramos Andinos. In: Libro de Investigación Del Proyecto Páramo Andino, Condesan, Universidad de Amsterdam, Universidad de Wisconsin, Quito, 2012.

Cuesta, F. and De Bievre, B.: Temperate grasslands of South America, Hohhot - China,2008, 41 pp.

Cuesta, F., Sevink, J., Llambí, L. D., De Bièvre, B., and Posner, J. (Eds.): Avances en investigación para la conservación de los páramos andinos, CONDESAN., 2013.

Cuo, L., Lettenmaier, D. P., Mattheussen, B. V., Storck, P., and Wiley, M.: Hydrologic predition for urban watersheds with the Distributed Hydrology-Soil-Vegetation Model, Hydrological Process, DOI: 10.1002/hyp.7023, 2008.

d'Arge, R., Limburg, K., Grasso, M., de Groot, R., Faber, S., O'Neill, R. V., Van den Belt, M., and al., e.: The Value of the World's Ecosystem Services and Natural Capital, 1997.

Dahlke, H. E., Lyon, S. W., Jansson, P., Karlin, T., and Rosqvist, G.: Isotopic investigation of runoff generation in a glacierized catchment in northern Sweden, Hydrol Process, 27, doi: 10.1002/hyp.9668, in press, 2012.

Daily, G. C., Polasky, S., Goldstein, J., Kareiva, P. M., Mooney, H. A., Pejchar, L., Ricketts, T. H., Salzman, J., and Shallenberger, R.: Ecosystem services in decision making: time to deliver, Frontier in Ecolog and the Environment, 7, 21-28, 2009.

Daly, C., Gibson, W. P., Taylor, G. H., Johnson, G. L., and Pasteris, P.: A Knowledge-Based Approach to the Statistical Mapping of Climate Climate Research, 22, 99-113, 2002.

Daniel, J.: Sampling Essentials. Practical Guidelines for Making Sampling Choices, SAGE Publications, Inc, Los Angeles, CA, 2012.

de Wit, C. T.: Philosophy and terminology. In: Leffelaar 3-9., 1993.

Dercon, G., Bossuyt, B., De Bievre, B., Cisneros, F., and Deckers, J.: Zonification agroecologica del Austro Ecuatoriano. In: Universidad de Cuenca, Cuenca, Ecuador, 1998.

Dercon, G., Govers, G., Poesen, J., Sanchez, H., Rombaut, K., Vandenbroeck, E., Loaiza, G., and Deckers, J.: Animal powered tillage erosion assessment in the southern Andes region of Ecuador, Geomorphology, 87, 4-15, 2007.

Di Vittorio, A. V., Anderson, R. S., White, J. D., Miller, N. L., and Running, S. W.: Development and optimization of an Agro-BGC ecosystem model for C4 perennial grasses Ecol. Model., 221, 2038-2053, 2010.

Diemer, M.: Leaf lifespans of high-elevation, a seasonal Andean shrub species in relation to leaf traits and leaf habit Global Ecological Biogeography, 7, 457–465, 1998.

Diemer, M.: Microclimatic convergence of high-elevation tropical paramo and temperature-zone alpine environments, Vegetation Science, 7, 821-830, 1996.

Doten, C. O., Bowling, L. C., Maurer, E. P., Lanini, J. S., and Lettenmaier, D. P.: A spatially distributed model for the dynamic prediction of sediment erosion and transport in mountainous forested watersheds Water Resour. Res., 42, 2006.

Downer, C. C.: The Mountain Tapir, Endangered 'flagship' species of the High Andes, Oryx, 30, 45-58, 1996.

Dufour, A., Gadallah, F., Wagner, H. H., Guisan, A., and Buttler, A.: Plant species richness and environmental heterogeneity in a mountain landscape: effects of variability and spatial configuration, Ecography, 29, 573-584, 2006.

Elsenbeer, H.: Hydrologic flowpaths in tropical rainforest soilscapes - a review, Hydrol. Process, 15, 1751-1759, 2001.

Elshorbagy, A., Corzo, G., Srinivasulu, S., and Solomatine, D. P.: Experimental investigation of the predictive capabilities of data driven modeling techniques in hydrology - Part 1: Concepts and methodology, Hydrol. Earth Syst. Sci, 14, 1931–1941, 2010.

Engstrom, R. and Hope, A.: Parameter sensitvity of the Arctic Biome-BGC model for estimating evapotranspiration in the Arctic coastal plain Arct. Antarct. Alp. Res., 43, 380-388, 2011.

Eswaran, H., van den Berg, E., and Reich, P.: Organic carbon in soils of the world Soil Science Society of America Journal, 57, 192–194, 1993.

FAO: Grassland carbon sequestration: management, policy and economics, Rome2010.

Farley, K.: Grasslands to Tree Plantations: Forest Transition in the Andes of Ecuador, Annals of the Association of American Geographers, 97, 755-771, 2007.

Farley, K. A., Anderson, W. G., Bremer, L. L., and Harden, C. P.: Compensation for Ecosystem Services: An Evaluation of Efforts to Achieve Conservation and Development in Ecuadorian Páramo Grasslands Environmental Conservation, 38, 393–405, 2011.

Farley, K. A., Bremer, L. L., Harden, C. P., and Hartsig, J.: Changes in carbon storage under alternative land uses in biodiverse Andean grasslands: implications for payment for ecosystem services, Conservation Letters, 6, 21-27, 2013.

Farley, K. A., Kelly, E. F., and Hofstede, R.: Soil organic carbon and water retention after conversion of grasslands to pine plantations in the Ecuadorian Andes., Ecosystems, 7, 729–739, 2004.

Farquhar, G. D., von Caemmerer, S., and Berry, J. A.: A biogeochemical model of photosynthetic CO_2 assimilation in leaves of C_3 species., Planta, 149, 78-90, 1980.

Favier, V., Coudrain, A., Cadier, E., Francou, B., Ayabaca, E., Maisincho, L., Praderio, E., Villacis, M., and Wagnon, P.: Evidence of groundwater flow on Antizana ice-covered volcano, Ecuador/Mise en évidence d'écoulements souterrains sur le volcan englacé Antizana, Equateur, Hydrolog. Sci, 53, 278-291, 2008.

Favier, V., Wagnon, P., Chazarin, J. P., Maisincho, L., and Coudrain, A.: One-year measurements of surface heat budget on the ablation zone of Antizana Glacier 5, Ecuadorian Andes, Geophys. Res., 109, D18105, doi: 18110.11029/12003JD004359, 2004.

Fiala, K.: Estimation of annual increment of undreground plant biomass in a grassland community (*Polygalo-Nardetum*), Folia Geobotanica et Phytotaxonomica, 14, 1-10, 1979.

Fioretto, A., Di Nardo, C., Papa, S., and Fuggi, A.: Lignin and cellulose degradation and nitrogen dynamics during decomposition of three leaf litter species in a Mediterranean ecosystem, Soil Biology & Biochemistry, 37, 1083–1091, 2005.

Fischlin, A., Midgley, G. F., Price, J. T., Leemans, R., Gopal, B., Turley, C., Rounsevell, M. D. A., Dube, O. P., Tarazona, J., and Velichko, A. A. (Eds.): Ecosystems, their properties, goods, and services. Climate Change 2007: Impacts, Adaptation and Vulnerability. Contribution of Working Group II to the Fourth Assessment Report of the Intergovernmental Panel on Climate Change, M.L. Parry, O.F. Canziani, J.P. Palutikof, P.J. van der Linden and C.E. Hanson, Eds., Cambridge University Press, Cambridge, 211-272., 2007.

Foot, K. and Morgan, R. P. C.: The role of leaf inclination, leaf orientation and plant canopy architecture in soil particle detachment by raindrops, Earth Surface Process Landforms, 30, 1509-1520, 2005.

Francou, B.: Recesion de los glacieares en el Ecuador: una respuseta al cambio climatico. In: Montaña, Quito, 2007.

Francou, B., Caceres, B., Gomez, J., and Soruco, A.: Coherence of the glacier signal throughout the tropical Andes over the last decades, Bogota2007, 87-97.

Francou, B., Ramirez, E., Caceres, B., and Mendoza, J.: Glacier evolution in the tropical Andes during the last decades of the 20th century: Chacaltaya, Bolivia and Antizana, Ecuador, Ambio, 29, 416-422, 2000.

Galelli, S., Humphrey, G., Maier, H., Castelletti, A., Dandy, G., and Gibbs, M.: An evaluation framework for input variable selection algorithms for environmental data-driven models, Environmental Modelling and Software, 62, 33-51, 2014.

Gamier, B. J. and Ohmura, A.: A method of calculating the direct shortwave radiation income of slopes, Journal of Applied Meteorology, 7, 796-800, 1968.

Ganskopp, D. and Rose, J.: Bunchgrass basal area affects selection of plants by cattle, Journal of Range Manage, 45, 538-541, 1992.

Gardi, C., Angelini, M., Barceló, S., Comerma, J., Cruz Gaistardo, C., Encina Rojas, A., Jones, A., Krasilnikov, P., Mendonça Santos Brefin, M. L., Montanarella, L., Muñiz Ugarte, O., Schad, P., Vara Rodríguez, M. I., and Vargas, R. (Eds.): Atlas de suelos de América Latina y el Caribe, Oficina de Publicaciones de la Unión Europea, L-2995, 176 pp, Luxembourg, 2014.

Genereux, D.: Quantifying uncertainty in tracer-based hydrograph separations Water Resour. Res. , 34, 915–919, 1998.

Gerten, D., Schaphoff, S., Haberlandt, U., Lucht, W., and Sitch, S.: Terrestrial vegetation and water balance – Hydrological evaluation of a dynamic global vegetation model J. Hydrol., 286, 249 – 270, 2004.

Gilbert, R. O.: Statistical Methods for Environmental Pollution Monitoring, Wiley, New York, 1987.

Gill., R. A. and Jackson, R. B.: Global patterns of root turnover for terrestrial ecosystems, New Phytologist, 147, 13-31, 2000.

Goller, R., Wilcke, W., Leng, M. J., Tobschall, H. J., Wagner, K., Valarezo, C., and Zech, W.: Tracing water paths through small catchments under a tropical montane rain forest in south Ecuador by an oxygen isotope approach, J. Hydrology, 308, 67-80, 2005.

Gomez-Diaz, J. D., Etchevers-Barra, J. D., Monterroso-Rivas, A. I., Campo-Alvez, J., and Tinoco-Rueda, J. A.: Ecuaciones alométricas para estimar biomasa y carbono en *Quercus magnoliaefolia*, Revista Chapingo Serie Ciencias Forestales y del Ambiente, 17, 261-272, 2011.

Gómez Molina, E. and Little, A. V.: Geoecology of the Andes: The natural science basis for research planning. , Mountain Res. Developm. , 1, 115-144, 1981.

Gough, C. M.: Terrestrial Primary Production: Fuel for Life, Nature Education, 3, 28, 2011.

Greiber, T. and Schiele, S. (Eds.): Governance of Ecosystem services, Gland, Switzerland, 2011.

Haines-Yong, R. and Potschin, M.: CICES V4.3 - Revised report prepared following consultation on CICES Version 4, August - December 2012, Centre for Environmental Management, University of Nottingham, UK, 2013.

Haines-Yong, R. H. and Potschin, M. (Eds.): The links between biodiversity, ecosystem services and human well-being, Cambridge University Press, Cambridge, 2010.

Hall, M., Mothes, P., Aguilar, J., Bustillos, J., Ramon, P., Eissen, J. P., Monzier, M., Robin, C., Egred, J., Militzer, A., and Yepes, H.: Los peligros volcánicos asociados con el Antisana. In: Serie los peligros volcánicos en el Ecuador, No. 4, Corporación Editora Nacional / IG-EPN / IRD, Quito, 2012.

Hall, M. L. and Beate, B. (Eds.): El Volcanismo Plio-Cuaternario en Los Andes del Ecuador. El Paisaje Volcánico de la Sierra Ecuatoriana. Geomorfología, fenómenos volcánicos y recursos asociados., Corporación Editora Nacional/Colegio de Geógrafos del Ecuador; Estudios de Geografía, Quito, 1991.

Hamza, M. A. and Anderson, W. K.: Soil compaction in cropping systems, Soil & Tillage Research 82, 121–145, 2005.

Hansen, B. C. S., Rodbell, D. T., Seltzer, G. O., Leon, B., Young, K. R., and Abbot, M.: Late-glacial and Holocene vegetation history from two sites in the Western Cordillera of the southwestern Ecuador, Palaeogeography, Palaeoclimatology, Palaeoecology, 194, 79-108, 2003.

Harden, C. P.: Human impacts on headwater fluvial systems in the northern and central Andes Geomorphology 79, 249–263, 2006.

Haxeltine, A. and Prentice, I. C.: BIOME3: An equilibrium terrestrial biosphere model based on ecophysiological constraints, resource availability, and competition among plant functional types Global Biogeochemical Cycles, 10, 693-709, 1996.

Hedberg, I. and Hedberg, O.: Tropical-alpine life-forms of vascular plants Oikos, 33, 297–307, 1979.

Hedberg, O.: Alfroalpine vegetation compared to the Paramos. Convergent adaptation and divergent differentiations., In: Balslev, H. & Luteyn, J.L., eds. Academic Press, London, 1992.

Hedberg, O.: Features of afroalpine plant ecology, Acta Phytogeographica Suecica, 49, 1-144, 1964.

Hein, L., van Koppen, K., de Groot, R. S., and van Ierland, E. C.: Spatial Scales, Stakeholders and the Valuation of Ecosystem Services Ecological Economics 57, 209–228. doi:210.1016/j.ecolecon.2005.1004.1005, 2006.

Hickler, T., Smith, B., Sykes, M. T., Davis, M. B., Sugita, S., and Walker, K.: Using a generalized vegetation model to simulate vegetations dynamics in Northeastern USA, Ecology, 85, 519-530, 2004.

Hidy, D., Barcza, Z., Haszpra, L., Churkina, G., Pintér, K., and Nagye, Z.: Development of the Biome-BGC model for simulation of managed herbaceous ecosystems, Ecological Modelling 226, 99–119, 2012.

Hilbert, D. W. and Ostendorf, B.: The utility of artificial neural networks for modelling the distribution of vegetation in past, present and future climates, Ecological Modelling, 146, 311-327, 2001.

Hilt, N. and Fiedler, K.: Diversity and composition of Arctiidae moth ensembles along a successional gradient in the Ecuadorian Andes. , Diversity and Distributions, 11, 387–398, 2005.

Hindshaw, R. S., Tipper, E. T., Reynolds, B. C., Lemarchand, E., Wiederhold, J. G., Magnusson, J., Bernasconi, S. M., Kretzschmar, R., and Bourdon, B.: Hydrological control of stream water chemistry in a glacial catchment (Damma Glacier, Switzerland), Chem. Geol, 285, 215-230, 2011.

Hofstede, R.: Effects of burning and grazing on a Colombian páramo ecosystem Ph.D. Dissertation, University of Amsterdam, The Netherlands, 1995.

Hofstede, R.: La importancia hídrica del páramo y aspectos de su manejo, EcoPar, 1997.

Hofstede, R., Coppus, R., Mena-Vasconez, P., Segarra, P., Wolf, J., and Sevink, J.: El estado de conservacion de los paramos de pajonal en el Ecuador, Ecotropicos, 15, 3–18, 2002.

Hofstede, R., Segarra, P., and Mena, P. V.: Los Paramos del Mundo. Quito, Ecuador, Global Peatland Initiative/NC-IUCN/EcoCiencia, 2003.

Holm, S.: A simple sequentially rejective multiple test procedure. , Scand J Statistics, 6, 65-70, 1979.

Hribljan, J. A., Suárez, E., Heckman, K. A., Lilleskov, E. A., and Chimner, R. A.: Peatland Carbon Stocks and Accumulation Rates in the Ecuadorian Páramo, Wetlands Ecology and Management, 24, 113–127, 2016.

Hudson, M., Hagan, M. T., and Demuth, H. B.: Neural Network Toolbox, User's Guide, MATLAB, Mathworks, Inc, 2014.

Hugenschmidt, C., Ingwersen, J., Sangchan, W., Sukvanachaikul, Y., Duffner, A., Unhlenbrook, S., and Streck, T.: A three-component hydrograph separation based on geochemical tracers in a tropical mountainous headwater catchment in northern Thailand, Hydrol. Earth Syst. Sci., 18, 525-537, 2014.

Hungerford, R. D., Nemani, R. R., Running, S. W., and Coughlan, J. C.: MTCLIM: A mountain microclimate simulation model, USFS Int Res. Stn Research Paper, # INT-414, 1989.

Hunt, E. R., Piper, S. C., Nemani, R., Keeling, C. D., Otto, R. D., and Running, S. W.: Global net carbon exchange and intra-annual atmospheric CO_2 concentrations predicted by an ecosystem process model and three-dimensional atmospheric transport model. , Global Biogeochemical Cycles, 10, 431-456, 1996.

Huss, M., Farinotti, D., Bauder, A., and Funk, M.: MOdelling runoff from highly glacierized alpine drainage basins in a changing climate, Hydrol Process, 22, 3888-3902, 2008.

Huston, M. A. (Ed.): Biological diversity: the co-existence of species on changing landscapes, Cambridge University Press, Cambridge, UK, 1994.

IAEA: http://www-naweb.iaea.org/napc/ih/IHS_resources_gnip.html, last access: 1 December 2015.

IAEA/GNIP: Precipitation sampling guide, V2.02 September 2014.

IGM: Instituto Geográfico Militar (IGM) Ecuador. Topographic map scale 1:25000, 1990.

INEFAN.: Guia para los páramos del Sistema Nacional de Areas protegidas del Ecuador, Artes Gráficas, Quito - Ecuador, 1996.

IPCC: Climate Change 2013: The Physical Science Basis. Contribution of Working Group I to the Fifth Assessment Report of the Intergovernmental Panel on Climate Change [Stocker, T.F., D. Qin, G.-K. Plattner, M. Tignor, S.K. Allen, J. Boschung, A. Nauels, Y. Xia, V. Bex and P.M. Midgley (eds.)]. Cambridge University Press, Cambridge, United Kingdom and New York, NY, USA, 2013.

IPCC.: Climate change 2007 – impacts, adaptation and vulnerability Cambridge, 2007.

Janeau, J. L., Grellier, S., and Podwojewski, P.: Influence of rainfall interception by endemic plants versus short cycle crops on water infiltration in high altitude ecosystems of Ecuador, Hydrology Research, 46, 1008-1018, 2015.

Jansky, L., Ives, J. D., Furuyashiki, K., and Watanabe, T.: Global mountain research for sustainable development, Global Environmental Change, 12, 231–239, 2002.

Jensen, M. E. and Bourgeron, B. S. (Eds.): A guidebook for Integrated Ecological Assessments, Springer-Verlag, New York, 2001.

Jobbagy, E. G. and Jackson, R. B.: The vertical distribution of soil organic carbon and its relation to climate and vegetation, Ecological Applications, 10, 423-436, 2000.

Joos, F., Prentice, I. C., and Sitch, S.: Global warming feedbacks on terrestrial carbon uptake under the Intergovernamental Panel on Climate Change (IPCC) emissions scenarios, Global Biogeochem Cycles, 15, 891-907, 2001.

Jorgensen, P. M. and Leon, S. (Eds.): Catalogue of the Vascular Plants of Ecuador., Missouri Botanical Garden Press, St. Louis Missouri U.S.A., 1999.

Jørgensen, P. M. and Ulloa, C.: Seed plants of the high Andes of Ecuador: A checklist. Aarhus, Department of Systematic Botany, Aarhus University, AAU Report 34, Aarhus, 1994.

Jung M, Reichstein M, Margolis H.A., and al., e.: Global patterns of land-atmosphere fluxes of carbon dioxide, latent heat, and sensible heat derived from eddy covariance, satellite, and meteorological observations Journal Of Geophysical Research, 116, doi: 10.1029/2010jg001566, 2011.

Jung, M., Verstraete, M., Gobron, N., Reichstein, M., Papale, D., Bondeau, A., Robustelli, M., and Pinty, B.: Diagnostic assessment of European gross primary production, Glob Chang Biology, 14, 2349–2364, 2008.

Jung, M., Vetter, M., Herold, M., Churkina, G., Reichstein, M., Zaehle, S., Ciais, P., Viovy, N., Bondeau, A., Chen, Y., Trusilova, K., Feser, F., and Heimann, M.: Uncertainties of modeling gross primary productivity over Europe: A systematic study on the effects of using different drivers and terrestrial biosphere models, Global Biogeochemical Cycles, 21, 2007.

Kadovic, R., Belanovic, S., Obratov-Petkovic, D., Bjedov, I., Perovic, V., Andelic, M., Knezevic, M., and Rankovic, N.: Soil organic carbon storage in mountain grasslands of the lake Plteau at Mr. Durmitor in Montenegro. In: Bulletin of the Faculty of Forestry, 2012.

Kaplan, J. O.: Geophysical applications of vegetation modelling, PhD thesis, University of Lund, 128 pp. pp., 2001.

Karssenberg, D., Schmitz, O., Salamon, P., de Jong, K., and Bierkens, M. F. P.: A software framework for construction of process-based stochastic spatio-temporal models and data assimilation Environmental Modelling & Software, 25, 489-502, doi: 410.1016/j.envsoft.2009.1010.1004, 2010.

Kasabov, K.: Foundations of Neural Networks, Fuzzy Systems and Knowledge Engineering. MIT Press: Cambridge, 1996.

Kaser, G., Grosshauser, M., and Marzeion, B.: Contribution potential of glaciers to water availability in different climate regimes, P. Natl. Acad. Sci. USA, 107, 20223-20227, 2010.

Kaser, G. and Osmaston, H. (Eds.): Tropical Glaciers, Cambridge University Press, 2002.

Kendall, M. G.: Rank Correlation Methods,4th edition, Charles Griffin, London, 1975.

Kimball, J. S., Jones, L. A., Zhang, K., Heinsch, F. A., McDonald, K. C., and Oechel, W. C.: A satellite approach to estimate land-atmosphere CO_2 exchange for Boreal and Arctic biomes using MODIS and AMSR-E IEEE Transactions on Geoscience and Remote Sensing 47, 569-587, 2009.

Kimball, J. S., Keyser, A. R., Running, S. W., and Saatchi, S. S.: Regional assessment of boreal forest productivity using an ecological process model and remote sensing parameter maps Tree Physiol., 20, 761–775, 2000.

Kimball, J. S., Running, S. W., and Nemani, R.: An improved method for estimating surface humidity from daily minimum temperature, Agricultural and Forest Meteorology, 85, 87-98, 1997.

King, R. T.: Wildlife and man, NY Conservationist, 20, 8-11, 1966.

Klaus, J. and McDonnell, J. J.: Hydrograph separation using stable isotopes: Review and evaluation, J. Hydrology, 505, 47-64, 2013.

Kleier, C. and Rundel, P. W.: Microsite requirements, population structure and growth of the cushion plant *Azorella compacta* in the tropical Chilean Andes, Austral Ecology, 29, 461-470, 2004.

Koca, D., Smith, B., and Sykes, M. T.: Modelling regional climate change effects on potential natural ecosystems in Sweden, Climatic Change 78, 381-406, 2006.

Körner, C.: Alpine Plant Life: Functional Plant Ecology of High Mountain Ecosystems, 2nd ed. Springer-Verlag, Heidelberg, 2003.

Körner, C.: The use of 'altitude' in ecological research, Trends in Ecology and Evolution, 22, 569-574, 2007.

Korol, R. L., Milner, K. S., and Running, S. W.: Testing a mechanistic model for predicting stand and tree growth Forest Science 42, 139-153, 1996.

Koziel, S. and Leifsson, L. (Eds.): Surroagte-based modeling and optimization: Applications in engineering. S. Koziel and L. Leifsson, Springer, 2013.

Küper, W., Kreft, H., Nieder, J., Köster, N., and Barthlott, W.: Large-scale diversity patterns of vascular epiphytes in Neotropical montane rain forests. , Journal of Biogeography, 31, 1477–1487, 2004.

La Marche, J. and Lettenmaier, D. P.: Effects of forest roads on flood flows in the Des chutes river, Washington. , Earth Surface Processes and Landforms 26, 115 – 134, 2001.

Laegaard, S.: Influence of fire in the grass páramo vegetation of Ecuador., In: Balslev, H. & Luteyn, J. L. (eds) Paramo: An Andean Ecosystem under Human Influence. Academic Press, London, UK 1992.

Lange, H.: Modelling carbon dynamics in forest ecosystems using Biome-BGC. Greenhouse-gas budget of soils under changing climate and land use (BurnOut), COST 639, 2006-2010, Norwegian Forest and Landscape Institute, Norway, 2010.

Lange, O. L., Belnap, J., and Reichenberger, H.: Photosynthesis of the cyanobacterial soil-crust lichen Collema tenax from arid lands in siuthern Utah, USA: role of water content on light and temperature responses of CO_2 exchange Functional Ecology, 12, 195-202, 1998.

Lauer, W.: Ecoclimatological conditions of the páramo belt in the tropical high mountains, Mountain Research and Development, 1, 209-221, 1981.

Law, B. E., Arkebauer, T., Campbell, J. L., Chen, J., Sun, O., Schwartz, M., van Ingen, C., and Verma, S.: Terrestrial carbon observations: Protocols for vegetation sampling and data submission. GTOS (Ed.), Rome, 2008.

Legendre, P. and Legendre, L.: Numerical Ecology. 2nd edn, Elsevier Science B.V., Amsterdam, The Netherlands, 1998.

Lemon, J.: Plotrix: a package ini the red light district of R, R-News, 6, 8-12, 2006.

Levin, S. A.: Ecosystems and the biosphere as complex adaptive systems, Ecosystems, 1, 431-436, 1998.

Li, H., Arias, M., Blauw, A., Los, H., Mynett, A. E., and Peters, S.: Enhancing generic ecological model for short-term prediction of Southern North Sea algal dynamics with remote sensing images, Ecological Modelling, 221, 2435-2446, 2010.

Li, H., Corzo Perez, G., Martinez, C., and Mynett, A. E.: Self-Learning Cellular Automata for Forecasting Precipitation from Radar Images, Hydrologic Engineering, 18, 206–211, 2013.

Li, J., Cheng, J., Shi, J., and Huang, F.: Brief introduction of back propagation (BP) neural network algorithm and its improvement, Advances in CSIE, 2, 553-558, 2012.

Linderman, M., Liu, J., Qi, J., An, L., Ouyang, Z., Yang, J., and Tan, Y.: Using artificial neural networks to map the spatial distribution of understorey bamboo from remote sensing data, International Journal of Remote Sensing, 25, 1685-1700, 2004.

Line, M. A. and Loutit, M. W.: Studies on non-symbiotic nitrogen fixation in New Zealand tussock-grassland soils, New Zealand Journal of Agricultural Research, 16, 87-94, 1973.

Liu, S., Costanza, R., Farber, S., and Troy, A.: Valuing Ecosystem Services, Annals of the New York Academy of Sciences 1185, 54–78, 2010.

Luteyn, J. L.: Páramos, a checklist of plant diversity, geographical distribution and botanical literature, New York Botanical Garden Press, New York, 1999.

Luteyn, J. L. and Balslev, H. (Eds.): Páramo: An Andean Ecosystem under Human Influence, Academic Press London (UK), London, 1992.

Madriñán, S., Cortés, A. J., and Richardson, J. E.: Páramo is the world's fastest evolving and coolest biodiversity hotspot, Frontiers in Genetics, 4, 192, 2013.

MAE.: (Ministerio del Ambiente del Ecuador). Acuerdo Ministerial No.115, MAE, Quito, Ecuador, 2009.

Manciati, C., Cadier, E., Galarraga-Sanchez, R., and Taupin, J. D.: South America Tropical glaciers retreat and temperature evolution in the Ecuadorian Andes. Case of Antisana volcano. In: Revista Peruana Geo-Atmosferica RPGA, Servicio Nacional de Metereologia e Hidrologia del Peru, Peru, 2011.

Mann, H. B.: Non-parametric tests against trend, Econometrica, 13, 163-171, 1945.

Marechal, J., Ladouche, B., Batiot-Guilhe, C., and Seidel, J.: Application of End-Member Mixing Analysis to karst hydrogeology, American Geophysical Union, Fall Meeting 2013, abstract #H23G-1356, 2013.

Marengo, J., Pabon, J., Diaz, A., Rosas, G., Avalos, G., Montealegre, E., Villacis, M., Solmana, S., and Rojas, M. (Eds.): Climate change: evidence and future scenarios for the Andean region Paris-France. In press, 2010.

Maskey, S.: HyKit – A Tool for Grid-based Interpolation of Hydrological Variables, User's Guide (Version 1.3). In: UNESCO-IHE Institute for Water Education, Delft, The Netherlands., 2013.

McGee, M.: http://co2now.org/, last access: December 12th 2014.

McGuire, A. D., Sitch, S., Clein, J. S., Dargaville, R., Esser, G., Foley, J., Heimann, M., Joos, F., Kaplan, J., Kicklighter, D. W., Meier, R. A., Melillo, J. M., Moore, B., Prentice, C., Ramankutty, N., Reichenau, T., Schloss, A., Tian, H., Williams, L. J., and Wittenberg, U.: Carbon balance of the terrestrial biosphere in the twentieth century: Analyses of CO2, climate and land use effects with four process-based ecosystem models, Global Biogeochem Cycles, 15, 183-206, 2001.

Medina, G. and Mena-Vásconez, P. (Eds.): Los Páramos en el Ecuador. En: Los Páramos del Ecuador, Proyecto Páramo y Abya Yala, Quito, 2001.

Mena, P., Josse, C., and Medina, G.: El páramo como fuente de recursos hídricos, Serie Páramo 3, 2000.

Mena, P. V.: Formas de vida de las plantas vasculares del páramo de El Angel y comparación con trabajos similares realizados en el cinturón afroalpino., Licenciatura, Departamento de Ciencias Biológicas, Pontificia Universidad Católica del Ecuador, Quito, 1984.

Mena, S. P.: Evolución de la dinámica de los escurrimientos en zonas de alta montaña: caso del Volcán Antisana, Civil Engineer, Facultad de Ingeniería Civil y Ambiental, Escuela Politécnica Nacional, Quito, 2010.

Milesi, C., Running, S. W., Elvidge, C. D., Dietz, J. B., Tuttle, B. T., and Nemani, R. R.: Mapping and Modeling the Biogeochemical Cycling of Turf Grasses in the United States, Environmental Management, 36, 426-438, 2005.

Miller, P. A., Giesecke, T., Hickler, T., Bradshaw, R. H. W., Smith, B., Seppä, H., Valdes, P. J., and Sykes, M. T.: Exploring climatic and biotic controls on Holocene vegetation change in Fennoscandia, Journal of Ecology, 96, 247-259, 2008.

Minaya, V., Corzo, G., Romero-Saltos, H., van der Kwast, J., Lantinga, E., Galarraga-Sanchez, R., and Mynett, A. E.: Altitudinal analysis of carbon stocks in the Antisana *páramo*, Ecuadorian Andes, Plant Ecology, 9, 553-563, 2015a.

Minaya, V., Corzo, G., Van der Kwast, J., Galarraga-Sanchez, R., and Mynett, A. E.: Classification and multivariate analysis of differences in gross primary production at different elevations using BIOME-BGC in the páramos; Ecuadorian Andean Region. In: Revista de Matemática: Teoría y aplicaciones Vol.22, No.2, 2, CIMPA, San Jose - Costa Rica. ISSN: 1409-2433 (print), 2215-3373 (online), 2015b.

Minaya, V., Corzo, G., van der Kwast, J., and Mynett, A. E.: Simulating gross primary production and stand hydrological processes of páramo grasslands in the Ecuadorian Andean Region using BIOME-BGC model, Soil Science, 181, 335-346, 2016.

Mitchell, T. M.: Machine Learning, McGraw-Hill, Singapore, 1997.

Molau, U.: Mountain biodiversity patterns at low and high latitudes, Ambio, 24-28 pp., 2004.

Molina, A., Govers, G., Vanacker, V., Poesen, J., Zeelmaekers, E., and Cisneros, F.: Runoff generation in a degraded Andean ecosystem: Interaction of vegetation cover and land use Catena 71, 357–370, 2007.

Monasterio, M.: Las formaciones vegetales de los páramos de Venezuela. In: Monasterio M (ed) Estudios ecológicos en los páramos andinos: 93-158. Universidad de Los Andes, Mérida, Venezuela., 1980.

Monasterio, M. and Sarmiento, L.: Adaptive radiation of Espeletia in the cold Andean tropics, Trends in Ecology and Evolution, 6, 387 - 391, 1991.

Monteith, J. L.: Principles of environmental physics, Elsevier, New York, 1973.

Moorcroft, P. R.: How close are we to a predictive scienced of the biosphere?, Trends in Ecology and Evolution, 21, 400-407, 2006.

Morales, P., Sykes, M. T., Prentice, I. C., Smith, P., Smith, B., Bugmann, H., Zierl, B., Friedlingstein, P., Viovy, N., Sabate, S., Sanchez, A., Pla, E., Gracia, C. A., Sitch, S., Arneth, A., and Ogee, J.: Comparing and evaluating process-based ecosystem model predictions of carbon and water fluxes in major European forest biomes Global Change Biology, 11, 2211-2233, 2005.

Mul, M. L., Mutiibwa, R. K., Unhlenbrook, S., and Savenije, H. H. G.: Hydrograph separation using hydrochemical tracers in the Mankanya catchment, Tanzania, Phys. Chem. Earth, 33, 151-156, 2008.

Munyaneza, O., Wenninger, J., and Unhlenbrook, S.: Identification of runoff generation processes using hydrometric and tracer methods in a meso-scale catchment in Rwanda, Hydrol Earth Syst Sci, 16, 1991-2004, 2012.

Myers, N., Mittermeier, R. A., Mittermeier, C. G., Fonseca, G. A. B., and Kent, J.: Biodiversity hotspots for conservation priorities, Nature, 403, 853-858, 2000.

Nagy, L. and Grabherr, G.: The biology of alpine habitats, Oxford University. Press, 2009.

Nagy, Z., Barcza, Z., Horváth, L., Balogh, J., Hagyó, A., Káposztás, N., Grosz, B., Machon, A., and Pintér, K. (Eds.): Measurements and estimations of biosphere-atmosphere exchange of greenhouse gases - grasslands., Springer, Dordrecht - Heidelberg - London - New York, pp 91-120, 2010.

Naveda-Rodríguez, A., Vargas, F. H., Kohn, S., and Zapata-Ríos, G.: Andean Condor (Vultur Gryphus) in Ecuador Geographic Distribution, Population Size and Extinction Risk PloS One, 11, e0151827., 2016.

Nelson, M. L., Rhoades, C. C., and Dwire, K. A.: Influence of bedrock geology on water chemistry of slope wetlands and headwater streams in the southern Rocky Mountains, Wetlands, 31, 251-261, 2011.

O' Callaghan, J. F. and Mark, D. M.: The extraction of drainage networks from digital elevation data, Comput. Vis. Graph. Image Process, 28, 328-344, 1984.

O'Connor, K. F., Nordmeyer, A. H., and Svavarsdóttir, K.: Changes in biomass and soil nutrient pools of tall tussock grassland in New Zealand In: Arnalds, O.; Archer, S. (Eds). Case studies of rangeland desertification. Proceedings from an international workshop in Iceland. Rala Report No. 200. Agricultural Research Institute, Reykjavik, 1999.

Odland, A.: Interpretation of altitudinal gradients in South Central Norway based on vascular plants as environmental indicators. , Ecological indicators, 9, 409-421, 2009.

Odum, E. P. (Ed.): Fundamentals of Ecology (3rd edn), W.B.Saunders, Philadelphia, USA, 1971.

Ojima, D. S., Parton, W. J., Schimel, D. S., Scurlock, J. M. O., and Kittel, T. G. F.: Modeling the effects of climatic and CO_2 changes on grassland storage of soil C, Water, Air, and Soil Pollution 70, 643-657, 1993.

Oksanen, J., Blanchet, F. G., Kindt, R., Legendre, P., Minchin, P., O'Hara, R. B., Simpson, G. L., Solymos, P., Stevens, M. H., and Wagner, H. (Eds.): Vegan: Community Ecology Package; R Package version 2.0-0, University of Oulu: Oulu - Finland. Available online: http://cran.r-project.org, 2011.

Ollinger, S. V., Aber, J. D., Reich, P. B., and Freuder, R. J.: Interactive effects of nitrogen deposition, tropospheric ozone, elevated CO_2 and land use history on the carbon dynamics of northern hardwood forests Glob. Change Biol. , 8, 545–562, 2002.

Osanai, Y., Flittner, A., Janes, J. K., Theobald, P., Pendall, E., Newton, P. C. D., and Hovenden, M. J.: Decomposition and nitrogen transformation rates in a temperate grassland vary among co-occurring plant species, Plant Soil, 350, 365-378, 2012.

Ott, B. and Uhlenbrook, S.: Quantifying the impact of land-use changes at the event and seasonal time scale using a process-oriented catchment model Hydrology and Earth System Sciences, 8, 62-78, 2004.

Papale, D. and Valentini, R.: A new assessment of European forests carbon exchanges by eddy fluxes and artificial neural network spatialization, Global Change Biology, 9, 525-535, 2003.

Parton, W. J., Schimel, D. S., Cole, C. V., and Ojima, D. S.: Analysis of factors controlling soil organic matter levels in grasslands Soil Science Society of America Journal, 51, 1173–1179, 1987.

Paruelo, J. M., Jobbagy, E. G., Sala, O. E., Lauenroth, W. K., and Burke, I. C.: Functional and structural convergence of temperate grassland and shrubland ecosystems, Ecological Applications 8, 194–206, 1997.

Paruelo, J. M. and Tomasel, F.: Prediction of functional characteristics of ecosystems: a comparison of artificial neural networks and regression models, Ecological Modelling, 98, 173-186, 1997.

Passioura, J. B.: Simulation Models: Science, Snake oil, Education, or Engineering? , Agronomy Journal 88, 690-694, 1996.

Paul, E. A.: Dynamics of soil organic matter, Plant and soil, 76, 275-285, 1984.

Pearce, A. J., Stewart, M. K., and Sklash, M. G.: Storm Runoff Generation in Humid Headwater Catchments 1. Where Does the Water Come From? , Water Resour Res, 22, 1263-1272, 1986.

Pfafstetter: Classification of hydrographic basins: coding methodology, unpublished manuscript, DNOS. August 18, 1989. Rio de Janeiro; translated by I. P. Vetdin, US Bureau of Reclamation. Brasilia, Brazil,September 5.1991, 1989.

Phoenix, G. K., Hicks, W. K., Cinderby, S., Kuylenstierna, J. C. I., Stock, W. D., Dentener, F. J., Giller, K. E., Austin, A. T., Lefroy, R. D. B., Gimeno, B. S., Ashmore, M. R., and Ineson, P.: Atmospheric nitrogen deposition in world biodiversity hotspots: the need for a greater global perspective in assessing N deposition impacts, Global Change Biology, 12, 470-476, 2006.

Pietsch, S. A., Hasenauer, H., and Thornton, P. E.: BGC-model parameters for tree species growing in central European forests., Forest Ecology and Management, 211, 264-295, 2005.

Pizarro, R., Araya, S., Jordan, C., Farıas, C., Flores, J. P., and Bro, P. B.: The effects of changes in vegetative cover on river flows in the Purapel river basin of central Chile Journal of Hydrology 327, 249–257, 2006.

Podwojewski, P.: Los Suelos de Las Altas Tierras Andinas: Los Páramos Del Ecuador, Bol Soc Ecuator Cie Suelo 18, 14, 1999.

Podwojewski, P., Poulenard, J., Zambrana, T., and Hofstede, R.: Overgrazing effects on vegetation cover and properties of volcanic ash soil in the páramo of Llangahua and La Esperanza (Tungurahua, Ecuador) Soil Use and Management, 18, 45–55, 2002.

Potschin, M. B. and Haines-Yong, R. H.: Ecosystem services: Exploring a geographical perspective, Progress in Physical Geography, 35, 575-594, 2011.

Poulenard, J., Bartoli, F., and Burtin, G.: Shrinkage and drainage in aggregates of volcanic soils: a new approach combining mercury porosimetry and vacuum drying kinetics, European Journal of Soil Science, 53, 563-574, 2002.

Poulenard, J., Podwojewski, P., and Herbillon, A. J.: Characteristics of non-allophanic andisols with hydric properties from the Ecuadorian páramos Geoderma, 117, 267–281, 2003.

Poulenard, J., Podwojewski, P., Janeau, J. L., and Collinet, J.: Runoff and soil erosion under rainfall simulation of andisols from the Ecuadorian paramo: Effect of tillage and burning Catena, 45, 185–207, 2001.

Prentice, I. C., Farquhar, G. D., Fasham, M. J. R., and al., E. (Eds.): The carbon cycle and the atmospheric carbon dioxide, Cambridge University Press, Cambridge, 2001.

Prentice, I. C., Heimann, M., and Sitch, S.: The carbon balance of the terrestrial biosphere: Ecosystem models and atmospheric observations, Ecological Applications, 10, 1553-1573, 2000.

Quillet, A., Peng, C., and Garneau, M.: Toward dynamic global vegetation models forsimulating vegetation–climate interactions andfeedbacks: recent developments, limitations, andfuture challenges, Environmental Reviews, 18, 333-353, 2010.

Quinlan, J. R.: Learning with continuous classes. , pp. 343-348, Singapore 1992.

R Development Core Team.: http://www.R-project.org/.

Rada, F., Garcia-Nuñez, C., Boero, C., Gallardo, M., Hillal, M., Gonzalez, J., Prado, F., Liberman-Cruz, M., and Azocar, A.: Low temperature resistance in *Polylepis tarapacana*, a tree growing at the highest altitudes in the world, Plant Cell Environment, 24, 377-381, 2001.

Rahman, M. M., Tsukamotob, J., Rahman, M. M., Yoneyama, A., and Mostafa, K. M.: Lignin and its effects on litter decomposition in forest ecosystems Chemistry and Ecology, 29, 540-553, 2013.

Ralp, C. P.: Observations on Azorella Compacta (*Umbelliferae*), atropical Andean cushion plant, Biotropica, 10, 62-67, 1978.

Ramsay, P. M. (Ed.): The Ecology of Volcán Chiles: high-altitude ecosystems on the Ecuador-Colombia border, Plymouth: Pebble & Shell 2001.

Ramsay, P. M.: The páramo vegetation of Ecuador: The community ecology, dynamics and productivity of tropical grasslands in the Andes Ph.D. thesis, [PhD thesis]. University of Wales, Bangor, 1992.

Ramsay, P. M. and Oxley, E. R. B.: The growth form composition of plant communities in the ecuadorian páramos, Plant Ecology, 131, 173-192, 1997.

Randerson, J. T., Chapin, F. S. I., Harden, J. W., Neff, J. W., and Harmon, M. E.: Net ecosystem production: a comprehensive measure of net carbon accummulation by ecosystems, Ecological Applications, 12, 937-947, 2002.

Rangel, O., Diaz, S., Jaramillo, R., and Salamanca, S.: Lista del material herborizado en el Transecto del Parque Los Nevados (*Pteridophyta-*

Spermatophyta), In: Van der Hammen, T., Pinto, P., Perez, A (eds.). La Cordillera Central Colombiana. Transecto Parque Los Nevados: Introducción y datos iniciales. *Estudios de ecosistemas tropandinos* 1: 174-205. J. Cramer, Vaduz., 1983.

Recknagel, F., French, M., Harkonen, P., and Yabunaka, K.: Artificial neural network approach for modelling and prediction of algal blooms, Ecological Modelling, 96, 11-28, 1997.

Regis, R. and Shoemaker, C.: Combining radial basis function surrogates and dynamic coordinate search in high-dimensional expensive black-box optimization, Engineering Optimization, 45, 529-555, 2013.

Ricardi, M., Gaviria, J., and Estrada, J.: La flora del superpáramo venezolano y sus relaciones fitogeográficas a lo largo de Los Andes, Plántula (Venezuela), 1, 171-187, 1997.

Roa-García, M. C., Brown, S., Schreier, H., and Lavkulich, L. M.: The role of land use and soils in regulating water flow in small headwater catchments of the Andes, Water Resour Res, 47, W05510, DOI: 05510.01029/02010WR009582, 2011.

Rodbell, D. T., Bagnato, S., Nebolini, J. C., Seltzer, G. O., and Abbott, M. B.: A late glacial–holocene tephrochronology for glacial lakes in southern Ecuador Quaternary Research, 57, 343–354, 2002.

Roderstein, M., Hertel, D., and Leuschner, C.: Above- and below-ground litter production in three tropical montane forests in southern Ecuador Journal of Tropical Ecology, 21, 483–492, 2005.

Ronquillo., J. C.: Guia de plantas del páramo de Papallacta. Reserva Ecologica Cayambe-Coca, Sendero "El Agua y la Vida", Fundacion Ecologica Rumicocha, Papallacta, Ecuador, 2010.

Rozanski, K., Araguás-Araguás, L., and Gonfiantini, R.: Isotopic patterns in modern global precipitation, Geophys. Monogr. Ser., 78, 1-36, 1993.

Rozanski, K., Levin, I., Stock, J., Guevara, R. E., and Rubio, F.: Atmospheric $^{14}CO_2$ variations in the equatorial region, Glasgow1995, 509-515.

Ruhl, J. B., Kraft, S. E., and Lant, C. L.: The Law and Policy of Ecosystem Services, Island Press, Washington, D.C., 2007.

Rumelhart, D. E., Hinton, G. E., and Williams, R. J.: Learning representations by back-propagating errors, Nature, 323, 1986.

Running, S. W. and Coughlan, J. C.: A general model of forest ecosystem processes for regional applications I. Hydrologic balance, canopy gas exchange and primary production processes, Ecological Modelling, 42, 125-154, 1988.

Running, S. W. and Gower, S. T.: FOREST-BGC, A general model of forest ecosystem processes for regional applications. II. Dynamic carbon allocation and nitrogen budgets Tree Physiol 9, 147-160, 1991.

Running, S. W. and Hunt, E. R. (Eds.): Generalization of a forest ecosystem process model for other biomes, BIOME-BGC, and an application for global-scale models. In: "Scaling physiological processes:leaf to globe", San Diego, CA, USA, pp.141-158, 1993.

Running, S. W., Nemani, R. R., and Hungerford, R. D.: Extrapolation of synoptic meteorological data in mountainous terrain and its use for simulating forest evaporation and photosynthesis, Journal of Forest. Res. , 17, 472-483, 1987.

Running, S. W., Thornton, P. E., Nemani, R., and Glassy, J. M. (Eds.): Global terrestrial gross and net primary productivity from the Earth Observing System, Springer-Verlag New York, 2000.

Sacramento, C. E., Pinheiro, D., da Silva, J., and Bayer, C.: Carbon sequestration in clay and silt fractions of Brazilian soils under conventional and no-tillage systems, Soils and Plant nutrition, 71, 292-301, 2014.

Sarmiento, L. and Bottner, P.: Carbon and nitrogen dynamics in two soils with different fallow times in the high tropical Andes: indications of fertility restoration Applied Soil Ecology, 19, 79-89, 2002.

Savanije, H. H. G.: The importance of interception and why we should delete the term evapotranspiration from our vocabulary, Hydrological Processes, 18, 1507-1511, 2004.

Savenije, H. H. G.: HESS Opinions: The art of hydrology, Hydrololgy and Earth System Sciences 13, 157–161, 2009.

Scanlon, T. M., Raffensperger, J. P., and Hornberger, G. M.: Modeling transport of dissolved silica in a forested headwater catchment: Implications for defining the hydrochemical response of observed flow pathways, Water Resour Res, 37, 1071-1082, 2001.

Scardi, M.: Artificial neural networks as empirical models for estimating phytoplankton production, Marine ecology progress series, 139, 289-299, 1996.

Schimel, D. S., Alves, D., Enting, I., Heimann, M., Joos, F., Raymond, D., and Wigley, T. (Eds.): CO$_2$ and the carbon cycle, in: Houghton, J.T. et al. (Eds.), Cambridge University Press, New York, pp.76-86, 1996.

Schlesinger, W. H. (Ed.): changes in soil carbon storage and associated properties with disturbance and recovery, Springer, New York, 1986.

Schroter, D. and et al.: Ecosystem Service Supply and Vulnerability to Global Change in Europe, Science, 310, 1333-1337, 2005.

Scott, D.: Methods of measuring growth in short tussocks, New Zealand Journal of Agricultural Research, 4, 282-285, 1961.

Scrucca, L.: GA: A Package for Genetic Algorithms in R, Journal of Statistical Software, 53, 1-37, http://www.jstatsoft.org/v53/i04/, 2012.

Shoji, S., Nanzyo, M., and Dahlgren, R.: Volcanic ash soils: genesis, properties and utilisation. Developments in Soil Science, Elsevier, Amsterdam, 1993.

Sieber, A. and Uhlenbrook, S.: Sensitivity analyses of a distributed catchment model to verify the model structure, Journal of Hydrology, 310, 216-235, 2005.

Sierra, R., Campos, F., and Chamberlin, J.: Assessing Biodiversity Conservation Priorities: Ecosystem Risk and Representativeness in Continental Ecuador, Landscape and Urban Planning 59, 95–110. doi:110.1016/S0169-2046(1002)00006-00003, 2002.

Sitch, S., Smith, B., Prentice, I. C., Arneth, A., Bondeau, A., Cramer, W., Kaplan, J. O., Levis, S., Lucht, W., Sykes, M. T., Thonicke, K., and Venevsky, S.: Evaluation of ecosystem dynamics, plant geography and terrestrial carbon cycling in the LPJ dynamic global vegetation model, Global Change Biol., 9, 161 – 185, 2003.

Six, J., Conant, R. T., Paul, E. A., and Paustian, K.: Stabilization mechanisms of soil organic matter: Implications for C-saturation of soils, Plant and Soil 241, 155–176, 2002.

Sklenár, P.: Decomposition of cellulose in the superpáramo of Ecuador Preslia, 70, 155–163, 1998.

Sklenar, P. and Jørgensen, P. M.: Distribution of páramo plants in Ecuador, Journal of Biogeography, 26, 681-691, 1999.

Sklenar, P. and Ramsay, P. M.: Diversity of Zonal Paramo Plant Communities in Ecuador, Diversity and Distributions, 7, 113-124, 2001.

Smith, B., Knorr, W., Widlowski, J. L., Pinty, B., and Gobron, N.: Combining remote sensing data with process modelling to monitor boreal conifer forest carbon balances, Forest Ecology and Management, 255, 3985-3994, 2008.

Smith, B., Prentice, I. C., and Sykes, M. T.: Representation of vegetation dynamics in the modelling of terrestrial ecosystems: comparing two contrasting approaches within European climate space Global Ecology and Biogeography, 10, 621-637, 2001.

Solomatine, D. and Dulal, K. N.: Model trees as an alternative to neural networks in rainfall-runoff modelling, Hydrological Siences, 48, 399-411, 2003.

Solomatine, D., See, L. M., and Abrahart, R. J.: Data-Driven Modelling: Concepts, Approaches and Experiences. In: Practical Hydroinformatics, al, R. J. A. e. (Ed.), Springer-Verlag Berlin Heidelberg, 2008.

Spain, A. V., Isbell, R. F., and Probert, M. E.: Soil organic matter, CSIR0, Melbourne Academic Press, London, UK, 1983.

Spehn, E. M., Liberman, M., and Korner, C. (Eds.): Land Use Change and Mountain Biodiversity, 2006.

Stern, M. J. and Guerrero, M. C.: Sucesión primaria en el Volcán Cotopaxi y sugerencias para el manejo de habitats frágiles dentro del Parque Nacional, Estudios sobre diversidad y ecología de plantas (ed. by R. Valencia and H. Balslev). Memorias del II Congreso de Botánica, PUCE, Quito, 1997.

Still, C. J., Foster, P. N., and Schneider, S. H.: Simulating the effects of climate change on tropical montane cloud forests., nature, 398, 1999.

Storck, P., Bowling, L., Wetherbee, P., and Lettenmaier, D. P.: Application of a GIS-based distributed hydrology model for prediction of forest harvest effects on peak stream flow in the Pacific Northwest., Hydrological Processes 12, 889 - 904, 1998.

Swift, L. W.: Algorithm for solar radiation on mountain slopes Water Resources Research, 12, 108-112, 1976.

Taiz, L. and Zeiger, E. (Eds.): Plant Physiology, 4th Ed, Sinauer Associates, Inc., Publishers, Sunderland, Massachusetts, 2006.

Tatarinov, F. A. and Cienciala, E.: Application of BIOME-BGC model to managed forests 1. Sensitivity analysis Forest Ecology and Management 237, 267-279, 2006.

Tatarinov, F. A. and Cienciala, E.: Long-term simulation of the effect of climate changes on the growth of main Central-European forest tree species, Ecol. Model., 220, 3081-3088, 2009.

Taylor, K. E.: Summarizing multiple aspects of model performance in a single diagram, Journal of Geophysical Research, 106, 7183-7192, 2001.

Thornton, P. E.: Biome-BGC Version 4.2 Final Release (C/C++). 2003.

Thornton, P. E.: Regional Ecosystem simulation: combining surface and satellite based observations to study linkages between terrestrial energy and mass budgets, PhD thesis, School of Forestry, University of Montana, Missoula, 280 p. pp., 1998.

Thornton, P. E., Hasenauer, H., and White, M. A.: Simultaneous estimation of daily solar radiation and humidity from observed temperature and precipitation: an application over complex terrain in Austria, Agricultural and Forest Meteorology, 104, 255-271, 2000.

Thornton, P. E., Law, B. E., Gholz, H. L., Clark, K. L., Falge, E., and co-authors, a.: Modeling and measuring the effects of disturbance history and climate on carbon and water budgets in evergreen needleaf forests, Agricultural Forest Meteorology, 113, 185-222, 2002.

Thornton, P. E. and Rosenbloom, N. A.: Ecosystem model spin-up: Estimating steady state conditions in a coupled terrestrial carbon and nitrogen cycle model, Ecological Modelling, 189, 25-48, 2005.

Thornton, P. E. and Running, S. W.: An improved algorithm for estimating incident daily solar radiation from measurements of temperature, humidity, and precipitation, Agricultural and Forest Meteorology, 93, 211-228, 1999.

Thornton, P. E., Running, S. W., and Hunt, E. R.: Biome-BGC: Terrestrial Ecosystem Process Model, Version 4.1.1. Data model. Available on-line [http://www.daac.ornl.gov] from Oak Ridge National Laboratory Distributed Active Archive Center. Oak Ridge, Tennessee, U.S.A., 2005.

Thoumi, G. and Hofstede, R.: Conserving Ecuador's paramos, the alpine tundra ecosystem of the Andes. In: special to mongabay.com, 2012.

Thyer, M., Beckers, J., Spittlehouse, D., Alila, Y., and Winkler, R.: Diagnosing a distributed hydrologic model for two high-elevation forested catchments based on detailed stand- and basin-scale data Water Resources Research 40, 2004.

Tilch, N., Unhlenbrook, S., and Leibundgut, C.: Regionalisation procedure for delineating hydrotopes in areas dominated by periglacial debris cover, Grundwasser, 2002/4, 206-216, 2002.

Tobon, C. and Gil Morales, E. G.: Capacidad de intercepcion de la niebla por la vegetacion de los paramos andinos, Avances en Recursos Hidraulicos, 15, ISSN-0121-5701, 2007.

Tonneijck, F. H., Jansen, B., Nierop, K. G. J., Verstraten, J. M., Sevink, J., and De Lange, L.: Towards understanding of carbon stocks and stabilization in volcanic ash soils in natural Andean ecosystems of northern Ecuador, European Journal of Soil Science, 61, 392-405, 2010.

Trusilova, K. and Churkina, G.: The terrestrial ecosystem model GBIOME-BGCv1, Max-Planck Institute for Biogeochemistry, Jena, Germany, 2008.

Trusilova, K., Trembath, J., and Churkina, G.: Parameter estimation and validation of the terrestrial ecosystem model BIOME-BGC using eddy-covariance flux measurements, Max Planck Institut fur Biogeochemie, Jena, 2009.

Tukey, J. W.: The Philosophy of Multiple Comparisons, Statistical Science, 6, 100-116, 1991.

Tupek, B., Zanchi, G., Verkerk, P. J., Churkina, G., Viovy, N., Hughes, J. K., and Lindner, M.: A comparison of alternative modelling approaches to evaluate the European forest carbon fluxes Forest Ecol. Manage., 260, 241-251, 2010.

Turner, D. P.: Scaling net ecosystem production and net biome production over a heterogeneous region in the western United States, Biogeosciences, 4, 597-612, 2007.

Ueyama, M., Harazono, Y., Kim, Y., and Tanaka, N.: Response of the carbon cycle in sub-arctic black spruce forests to climate change: Reduction of a carbon sink related to the sensitivity of heterotrophic respiration Agric. Forest Meteorol, 149, 582-602, 2009.

Uhlenbrook, S.: Untersuchung und Modellierung der Abflussbildung in einem mesoskaligen Einzugsgebiet, Freiburger Hydrologische Schriften, Band 10, Institut für Hydrologie, Universität Freiburg, Freiburg, 1999.

Uhlenbrook, S., Frey, M., Leibundgut, C., and Maloszewski, P.: Hydrograph separations in a mesoscale mountainous basin at event and seasonal timescales Water Resour Res, 38, 2002.

Uhlenbrook, S. and Leibundgut, C.: Process-oriented catchment modelling and multiple-response validation, Hydrol. Process, 16, 423-440, 2002.

Uhlenbrook, S., Roser, S., and Tilch, N.: Hydrological process representation at the meso-scale: the potential of a distributed, conceptual catchment model, Journal of Hydrology, 291, 278-296, 2004.

Uhlenbrook, S. and Sieber, A.: On the value of experimental data to reduce the prediction uncertainty of a process-oriented catchment model, Environmental Modelling & Software, 20, 19-32, 2005.

UN: Transforming our world: The 2030 agenda for sustainable development, 2015.

UNFCCC: Kyoto Protocol Reference Manual on accounting of Emissions and assigned amount, United Nations Framework Convention on Climate Change, 2008.

Urrutia, R. and Vuille, M.: Climate change projections for the tropical Andes using a regional climate model: temperature and precipitation simulations for the end of the 21st century Journal of Geophysical Research, 114, D02108, 2009.

van Keulen, H. and van Diepen, C. A.: Crop growth models and agroecological characterization, Paris. 1990.

Vanderbilt, K. L., White, C. S., Hopkins, O., and Craig, J. A.: Aboveground decomposition in arid environments: results of a long-term study in central New Mexico, Journal of Arid Environments, 72, 696–709, 2008.

Vargas, J. O., Premauer, J., and Cárdenas, C.: Efecto del Pastoreo Sobre la Estructura de la Vegetación en un Páramo Húmedo de Colombia, Ecotropicos 15, 35-50, 2002.

Verschot, L., Krug, T., Lasco, R. D., Ogle, S., and Raison, J.: IPCC Guidelines for National Greenhouse Gas Inventories, Agriculture, forestry and other land use. Volume 4, Chapter 6: Grassland. , 2006.

Verweij, P. A.: Spatial and temporal modelling of vegetation patterns: Burning and grazing in the páramo of Los Nevados National Park, Colombia., Ph.D. Dissertation. , University of Amsterdam, The Netherlands, Amsterdam, 1995.

Verweij, P. A. and Budde, P. E. (Eds.): Burning and grazing gradients in the paramo of Parque Los Nevados, Colobia: initial ordination analyses, Academic Press, London, 1992.

Vetter, M., Churkina, G., Jung, M., Reichstein, M., Zaehle, S., Bondeau, A., Chen, Y., Ciais, P., Feser, F., Freibauer, A., Geyer, R., Jones, C., Papale, D., Tenhunen, J., Tomelleri, E., Trusilova, K., Viovy, N., and Heimann, M.: Analyzing the causes and spatial pattern of the European 2003 carbon flux anomaly using seven models, Biogeosciences, 5, 561–583, 2008.

Villacis, M.: Ressources en eau glaciaire dans les Andes d'Equateur en relation avec les variations du climat: le cas du volcan Antisana, PhD Thesis UniversitéMontpellier II, Montpellier, France., 2008.

Villacis, M., Vimeux, F., and Taupin, J. D.: Analysis of the climate controls on the isotopic composition of precipitation ($\delta^{18}O$) at Nuevo Rocafuerte, 74.5°W, 0.9°S, 250 m, Ecuador, C.R. Geoscience, 340, 1-9, 2008.

Viviroli, D., Archer, D. R., Buytaert, W., Fowler, G., Greenwood, G. B., Hamlet, A. F., Huang, Y., Koboltschnig, G., Litaor, M. I., López-Moreno, J. I., Lorentz, S., Schadler, B., Schreier, H., Schwaiger, K., Vuillle, M., and Woods, R.: Climate change and mountain water resources: overview and recommendations for research, management and policy., Hydrol. Earth Syst. Sci., 15, 2011.

Vuille, M., Bradley, R. S., and Keimig, F.: Interannual climate variability in the Central Andes and its relation to tropical Pacific and Atlantic forcing, Geophys. Res.-Atmos., 105, 12447-12460, 2000.

Vuilleumier, F. and Monasterio, M. (Eds.): High altitude tropical biogeography, Oxford University Press, Oxford, 1986.

Walls, L. D.: The Passage to Cosmos: Alexander von Humboldt and the Shaping of America, University of Chicago Press, Chicago, 2009.

Walpole, M., Brown, C., Tierney, M., Mapendembe, A., Viglizzo, E., Goethals, P., Birge, T., and al., e.: Developing ecosystem service indicators: Experiences and lessons learned from sub-global assessments and other initiatives, Secretariat of the Convention on Biological Diversity, 58, 2011.

Walter, H.: Vegetation of the Earth in relation to climate and the eco-physiological conditions, Springer-Verlag, New York, 1973.

Wang, W., Ichiic, K., Hashimoto, H., Michaelisa, A. R., Thornton, P. E., Lawe, B. E., and Nemanib, R. R.: A hierarchical analysis of terrestrial ecosystem model Biome-BGC: Equilibrium analysis and model calibration, Ecological Modelling, 220, 2009-2023, 2009.

Wang, Y. and Witten, I. H.: Induction of model trees for predicting continuous classes, pp. 128-137, Prague1997.

Waring, H. R. and Running, S. W.: Forest Ecosystems. Analysis at multiples scales. 2nd edition, San Diego, pp.55, 2007.

Wenninger, J., Uhlenbrook, S., Lorentz, S., and Leibundgut, C.: Identification of runoff generation processes using combined hydrometric, tracer and geophysical methods in a headwater catchment in South Africa, Hydrological Sciences Journal, 53, 65-80, 2008.

Wesseling, C. G., Karssenberg, D., Burrough, P. A., and van Deursen, W. P. A.: Integrating dynamic environmental models in GIS: The development of a Dynamic Modelling language Transactions in GIS, 1, 40-48, 1996.

White, M. A., Thornton, P. E., and Running, S. W.: Parameterization and sensitivity analysis of the BIOME-BGC terrestrial ecosystem model: net primary production controls. , Earth Interactions 4, 1–85, 2000a.

White, P. S. and Pickett, S. T. A. (Eds.): Natural disturbance and patch dynamics: An introduction, pp 3-13. In: S. T.A. Pickett & P. S. White (eds) The Ecology of natural disturbance and patch dynamics. , New York, 1985.

White, R., Murray, S., and Rohweder, M.: Pylot analysis of global ecosystem: Grassland Ecosystems. Institute, W. R. (Ed.), Washington D.C., 2000b.

Wigmore, O. and Gao, J.: Spatiotemporal dynamics of a páramo ecosystem in the northern Ecuadorian Andes 1988-2007, Journal of Mountain Science, 11, 708-716, 2014.

Wigmosta, M. S., Nijssen, B., Storck, P., and Lettenmaier, D. P.: The Distributed Hydrology Soil Vegetation Model, In Mathematical Models of Small Watershed Hydrology and Applications, Water Resource Publications, 7-42, 2002.

Wigmosta, M. S., Vail, L., and Lettenmaier, D. P.: A distributed hydrology-vegetation model for complex terrain, Wat. Resour. Res., 30, 1665-1679, 1994.

Wilson, J.: "Alexander von Humboldt: A Chronology from 1769 to 1859" in Personal Narrative of A Journey to the Equinoctial Regions of the New Continent by Alexander von Humboldt Penguin Books. p. lxvii., London, 1995.

Windhorst, D., Waltz, T., Timbe, E., Frede, H. G., and Breuer, L.: Impact of elevation and weather patterns on the isotopic composition of precipitation in a tropical montane rainforest, Hydrol Earth Syst Sci, 17, 409-419, 2013.

Wissmeier, L.: Implementation of distributed solute transport into the catchment model TAC^d and event based simulations using oxygen-18, PhD, Institut fur Hydrologie, Universitat Freiburg, 154 pp., 2005.

Wissmeier, L. and Uhlenbrook, S.: Distributed, high-resolution modelling of 18O signals in a meso-scale catchment, Journal of Hydrology, 332, 497-510, 2007.

Witten, I. H. and Frank, E.: Data mining: Practical machine learning tools and techniques with Java implementations, Morgan Kaufmannp., 2000.

Wolf, A., Callaghan, T. V., and Larson, K.: Future changes in vegetation and ecosystem function of the Barents Region, Climate Change, 87, 51-73, 2008.

Wullschleger, S. D.: Biochemical limitations to carbon assimilation in C_3 plants - A retrospective analysis of the *A/Ci* curves from 109 species, Journal of Experimental Botany, 44, 907-920, 1993.

Xiao, J., Ollinger, S. V., Frolking, S., Hurtt, G. C., Hollinger, D. Y., Davis, K. J., Pane, Y., Zhang, X., Deng, F., Chen, J., Baldocchi, D. D., Law, B. E., Arain, M. A., Desai, A. R., Richardson, A. D., Sunn, G., Amiro, B., Margolis, H., Gu, L., Scott, R. L., Blanken, P. D., and Suyker, A. E.: Data-driven diagnostics of terrestrial carbon dynamics over North America, Agricultural and Forest Meteorology, 197, 142-157, 2014.

Yimer, F., Ledin, S., and Abdelkadir, A.: Changes in soil organic carbon and total nitrogen contents in three adjacent land use types in the Bale Mountains, south-eastern highlands of Ethiopia, Forest Ecology and Management, 242, 337-342, 2007.

Zabala, R. and Falconi, C.: El ecosistema paramo y potenciales pagos por conservacion de carbono, Santo Domingo - Ecuador 2010.

Zapata-Ríos, G. and Lyn, C. B.: Altered Activity Patterns and Reduced Abundance of Native Mammals in Sites with Feral Dogs in the High Andes, Biological Conservation 193, 9-16, 2016.

Zhang, Z., Verbeke, L., Clercq, E., Ou, X., and Wulf, R.: Vegetation change detection using artificial neural networks with ancillary data in Xishuangbanna, Yunnan Province, China Chinese Science Bulletin, 52, 232-243, 2007.

Zimmerer, K. S. (Ed.): Mapping Mountains in Mapping Latin America: a Cartographic Reader, University of Chicago Press, Chicago, 2011.

Acronyms

ADF	Acid Detergent Fiber
ANN	Artificial Neural Networks
ANOVA	One-way ANalysis Of VAriance
AR	Acaulescent Rosettes
ASTER	Advanced Spaceborne Thermal Emission and Reflection Radiometer
BGC	BioGeochemical Cycles
CMIP	Coupled Model Intercomparison Project
CWFS	Center of World Food Studies
CU	CUshions
DDM	Data Driven Model
DEM	Digital Elevation Model
DIM	Dimensionless
DGVM	Dynamic Global Vegetation Models
EC	Electrical Conductivity
EMMA	End Member Mixing Analysis
EPMAPS	Empresa Pública Metropolitana de Agua Potable y Saneamiento de Quito
EPN	Escuela Politécnica Nacional
GA	Genetic Algorithm
GBA	Grassland Basal Area
GIS	Geographical Information System
GPP	Gross Primary Production
GPS	Global Positioning System
GMWL	Global Meteoric Water Line
HBV	Hydrologiska Byråns Vattenbalansavdelning (model)
HU	Hydrological Units
IAEA	International Atomic Energy Agency
IBL	Instance-based learning
INAMHI	Instituto Nacional de Meteorología e Hidrología en Ecuador
INIGEMM	Instituto Nacional de Investigación Geológico Minero Metalúrgico
IPCC	Intergovernmental Panel on Climate Change

IRD	Institut de Recherche pour le Développement - Ecuador
LAI	Leaf Area Index
LANDIS	Landscape Model of Forest Dynamics
LMWL	Local Mean Water Line
LPJ	Lund-Potsdam-Jena (vegetation model)
MDS	MultiDimensional Scaling
MT	Model Trees
MT-CLIMB	Mountain Climate Model
NDVI	Normalized Difference Vegetation Index
NEP	Net Ecosystem Production
NPP	Net Primary Production
PERMANOVA	PERmutational Multivariate ANalysis Of VAriance
PFT	Plant Functional Type
PVC	Polymerizing Vinyl Chloride
QCA	Catholic University's Herbarium
QCNE	National Herbarium of Ecuador
RMSE	Root Mean Square Error
SDG	Sustainable Development Goal
SLA	Specific Leaf Area
SWR	Short Wave Radiation
TAC^D	Tracer Aided Catchment model, distributed
TU	TUssocks
VPD	Vapor Pressure Deficit
VSMOW	Vienna Standard Mean Ocean Water
WaSiM	Water balance Simulation Model
WOFOST	WOrld FOod STudies

Biography

Veronica Graciela Minaya Maldonado was born in Quito, Ecuador. She graduated as Civil Engineer in 2005 and later obtained a MSc in Hydric Resources and Water Science (2008), both from the Escuela Politecnica Nacional in Quito, Ecuador. She had some professional experience as Environmental Engineer and hydrologist with extensive fieldwork in the highlands and rainforest regions in Ecuador.

In 2010, she obtained her MSc in Environmental Sciences - specialization Limnology and Wetland Ecosystem, after attending modules in The Netherlands, Czech Republic and Austria. Her thesis was entitled "Land use change impacts on the benthic macro-invertebrate communities of streams in the head waters of Mara River, Kenya", it involved an extensive fieldwork in The Mara River Basin in Kenya that was fully founded by NUFFIC and USAID through Global Water for Sustainability Program (GLOWS). After her MSc, she did an internship at Deltares, where she got involved in different projects related to ecology, and several modelling applications.

The motivation to explore the ecohydrology of the *páramos* in her home country started many years ago during a field campaign. She was working as a consultant for a project to explore additional sources of water to cover the water demand of Quito by the year 2025. She was caught by fog and trapped in a situation that got her to experience the *páramo* unique features. She started her PhD research in the alpine grasslands in Ecuador since February 2012 at UNESCO - IHE in the Department of Water Science and Engineering. She received financial support from the Ecuadorian Government through SENESCYT (Secretaria Nacional de Educacion Superior, Ciencia, Tecnologia e Innovacion) and a complementary grant from the Dutch Ministry of Foreign Affairs (Netherlands Fellowship Programmes NFP).

She is actively involved in youth organizations, president of the IAHR - YPN (Young Professionals Network) in Delft, event leader and organizer of technical sessions and social activities for the young professionals that attended the 36th IAHR World Congress. In addition, she is a core member of the Water Youth Network, engagement and active participation in the 7th World Water Forum 2015, Stockholm World Water Week 2015 and Singapore International Water Week 2016. In general she is very interested in research development, and women & youth empowerment.

List of publications

Journal papers

Minaya, V., Corzo, G., Romero-Saltos, H., van der Kwast, J., Lantinga, E., Galarraga-Sanchez, R., and Mynett, A. E.: Altitudinal analysis of carbon stocks in the Antisana *páramo*, Ecuadorian Andes, Plant Ecology, 9, 5, 553-563, doi:10.1093/jpe/rtv073, 2015.

Minaya, V., Corzo, G., van der Kwast, J., and Mynett, A. E.: Simulating gross primary production and stand hydrological processes of *páramo* grasslands in the Ecuadorian Andean Region using BIOME-BGC model, Soil Science, 181, 7, 335-346, 2016.

Minaya V., Corzo G., Solomatine, D., and Mynett, A. E.: Data-driven techniques for modeling the gross primary production of the *páramo* vegetation using climate time-series data, application in the Ecuadorian Andean region, submitted to *Ecological Informatics*.

Minaya, V., Camacho, V., Weninger, J., Corzo, G., and Mynett, A. E.: Runoff generation of a combined glacier and *páramo* catchment within an Ecological Reserve in the Ecuadorian highlands, doi:10.5194/hess-2016-569, *Manuscript under review for Journal Hydrology and Earth System Sciences*, 2016.

Minaya V., Jianning, R., Mishra, P., Corzo, G., van der Kwast, J., Uhlenbrook, S., and Mynett, A. E.: Process-oriented hydrological representation of a *páramo* catchment; Ecuadorian Andean Region, Manuscript in preparation.

Minaya, V., Gonzalez-Angarita, A., Corzo, G., van der Kwast, J., and Mynett, A. E.: Ecosystem Service Indicators for Regulation and Maintenance type of Ecosystem Services derived from the *páramos*, Manuscript in preparation.

Other publications

Minaya, V., Corzo, G., Van der Kwast, J., Galarraga-Sanchez, R., and Mynett, A. E.: Classification and multivariate analysis of differences in gross primary production at different elevations using BIOME-BGC in the *páramos*; Ecuadorian Andean Region. In: Revista de Matemática: Teoría y aplicaciones Vol.22, No.2, 2, CIMPA, San Jose - Costa Rica. ISSN: 1409-2433, 2015.

Conference papers

Minaya V., Corzo G., van der Kwast J., Mynett A.: Altitudinal zonation of species distribution, carbon stocks and biomass quantification of the main growth forms of vegetation in the Ecuadorian alpine grassland. International Conference of Environment and Energy, Colombo - Sri Lanka, 2013.

Minaya V., Corzo G., Solomatine, D., Mynett A.: A data-driven technique for modeling the gross primary production of the *páramo* vegetation from climate time-series data; Ecuadorian Andean Region (Conference paper). 9th International Conference on Ecological Informatics (ICEI 2014), Nanjing - China, 2014.

Conference presentations and posters

Minaya V., van der Kwast J., Corzo G., Mynett A. The interrelation between water and vegetation using a coupled eco-hydrological model in the alpine grasslands (*páramo*) in the Ecuadorian Andean region. Course of Watershed Ecology and Biogeochemistry, Vindeln - Sweden, October 2013 [poster and oral presentation]

Minaya V., Corzo G., van der Kwast J., Mynett A. Altitudinal zonation of species distribution, carbon stocks and biomass quantification of the main growth forms of vegetation in the Ecuadorian alpine grassland. International Conference of Environment and Energy, Colombo - Sri Lanka, December 2013 [oral presentation].

Minaya V., Corzo G., van der Kwast J., Galarraga-Sanchez, R., Mynett A. Classification and multivariate error analysis of gross primary production simulation with BIOME-BGC in the *páramos*; Ecuadorian Andean Region, San Jose - Costa Rica, February 2014 [oral presentation].

Minaya V., Corzo G., van der Kwast J., Mynett A. Copula Multivariate analysis of Gross primary production and its hydro-environmental driver; A BIOME-BGC model applied to the Antisana *páramos*. European Geoscience Union (EGU), Vienna - Austria, April 2014 [poster presentation].

Minaya V., Corzo G., Solomatine, D., Mynett A. (2014) A data-driven technique for modeling the gross primary production of the *páramo* vegetation from climate time-series data; Ecuadorian Andean Region. 9th International Conference on Ecological Informatics (ICEI 2014), Nanjing - China, October 2014 [oral presentation]

Netherlands Research School for the
Socio-Economic and Natural Sciences of the Environment

D I P L O M A

For specialised PhD training

The Netherlands Research School for the
Socio-Economic and Natural Sciences of the Environment
(SENSE) declares that

Veronica Graciela Minaya Maldonado

born on 30 August 1981 in Quito, Ecuador

has successfully fulfilled all requirements of the
Educational Programme of SENSE.

Delft, 20 December 2016

the Chairman of the SENSE board

Prof. dr. Huub Rijnaarts

the SENSE Director of Education

Dr. Ad van Dommelen

The SENSE Research School has been accredited by the Royal Netherlands Academy of Arts and Sciences (KNAW)

K O N I N K L I J K E   N E D E R L A N D S E
A K A D E M I E   V A N   W E T E N S C H A P P E N

The SENSE Research School declares that Ms Veronica Minaya Maldonado has successfully fulfilled all requirements of the Educational PhD Programme of SENSE with a work load of 61.1 EC, including the following activities:

SENSE PhD Courses

o   Environmental research in context (2012)
o   Research in context activity: 'Co-organising the SENSE writing week' (2013)
o   The role of plants in Earth's climate, SENSE Summer Academy, Utrecht (2013)
o   SENSE writing week (2013)

Other PhD and Advanced MSc Courses

o   Hyper-temporal Earth observation data analysis for food security and biodiversity assessment, University of Twente (2012)
o   Watershed ecology and biogeochemistry, Swedish University of Agricultural Sciences (2013)
o   Data driven modelling and real-time control of water systems, UNESCO-IHE Delft (2014)

Management and Didactic Skills Training

o   President of the International Association for Hydro-Environment Engineering and Research Young Professional Network (IAHR -YPN) (2015)
o   Core member of the Water Youth Network (WYN) (2015)
o   Co-organising the course 'LateX', UNESCO-IHE Delft (2015)

Oral Presentations

o   *Analysis of biomass and carbon distribution in the Ecuadorian alpine grassland.* International Conference on Environment & Energy, 16-17 December 2013, Colombo, Sri Lanka
o   *Classification and multivariate error analysis of primary production simulation with BIOME-BGC in the paramos; Ecuadorian Andean Region.* XIX International Symposium of Mathematical Methods Applied to Sciences, 25-28 February 2014, San Jose, Costa Rica
o   *Analysis of biomass and carbon distribution in the Ecuadorian alpine grassland.* 9th International Conference on Ecological Informatics, 20-23 October 2014, Nanjing, China

SENSE Coordinator PhD Education

Dr. ing. Monique Gulickx

Printed and bound by CPI Group (UK) Ltd, Croydon, CR0 4YY

22/10/2024

01777614-0002